SpringerBriefs in Social Work

More information about this series at http://www.springer.com/series/13578

Sana Loue
Editor

Ethical Issues in Sandplay Therapy Practice and Research

 Springer

Editor
Sana Loue
Case Western Reserve University
Cleveland, Ohio
Ohio
USA

ISSN 2195-3104   ISSN 2195-3112 (electronic)
ISBN 978-3-319-14117-6   ISBN 978-3-319-14118-3 (eBook)
DOI 10.1007/978-3-319-14118-3

Library of Congress Control Number: 2015935434

Springer Cham Heidelberg New York Dordrecht London

Printed on acid-free paper

Springer is a brand of Springer International Publishing
Springer is part of Springer Science+Business Media (www.springer.com)

Preface

I first learned of sandplay therapy in 1986 when, as a client, I embarked on an inward journey to find my balance and my place in the world. In sandplay I encountered what seemed to be magical figures that spoke to me and from me that engaged with me and through me, serving as mediators between my inner and outer worlds.

Since that time, I have remained drawn to and captivated by sandplay as a modality, both in the roles of client and of therapist. It is a modality unlike any other—one that is nonverbal and nonrational, allowing us to reach a preverbal level of the psyche (Weinrib 1983, p. 1). Unlike many other therapeutic modalities, the sandplay therapist does not attempt to interpret to the client the sandplay scenes at the time they are made. Instead, he or she may elicit the client's comments or observations about what has been done in the sand. The therapist will use Jungian symbology and archetypal amplifications to understand the sandtrays that have been made. These interpretations, however, stand as hypotheses; they are not presented to the client at the time of the sandtray, to be subjected to affirmation, modification, or refutation.

Both the diversity and the universality of expression through sandplay allow the practicing therapist to learn and understand to a greater degree the complexities and wonderment of human nature, experience, and resilience. Our sharing of these understandings, through research, training, and practice with others, is vital if we are to be able to facilitate our clients' growth as they move forward in their journeys toward wholeness and, indeed, to know ourselves better and become better therapists.

As we move forward in our learning and our sharing, both as individual therapists and as members of professional organizations engaged with sandplay therapy, it is critical that we recognize and remain attuned to the ethical issues that arise. Many times there are no easy answers or solutions to the dilemmas that we confront—the child who desperately wishes to continue with therapy but whose parent refuses to allow it; the need for ongoing professional consultation with relatively few available therapists to provide it; maintaining one's role as therapist with a client who one sees on a regular basis through other activities; facilitating access to sandplay therapy for those potential clients who want it and could benefit from it, but whose funds and life situations do not permit them that access. Many of these issues are common across therapeutic modalities but offer particular challenges in the context of sandplay therapy.

Carina Conradie and Roxie Hanes address in Chap. 1 one of the threshold ethical issues related to sandplay practice: how to recognize one's own level of competence with sandplay practice and the need for continued training and competent supervision while in training. They ask the profound question, "How can we expect our client to experience this deep level of healing if we have very little experience of the sandplay process ourselves as practitioners?"

Jean Parkinson and Sana Loue continue to examine the ethical questions associated with supervision-consultation that is conducted electronically, whether through fax, e-mail, the Cloud, Skype, Oovoo, or electronic means. The use of electronic modalities for supervision and consultation has become increasingly common in sandplay therapy, due to the relative dearth in many geographical locations of sandplay therapists who are qualified to offer such services. Clearly, the therapists seeking the services and those providing them are doing so in an effort to improve the quality of care provided to clients and to heighten therapist competence. Even so, the use of electronic means to accomplish these purposes carries inherent risks to client confidentiality and privacy, demanding that we look further at how to best protect our clients.

Loue examines in Chap. 3 the ability of potential clients to access sandplay therapy from the vantage point of the ethical principle of distributive justice. Sandplay is a specialized modality of therapy; it would neither be desired by nor beneficial to every individual. It is no more elitist in terms of the cost of therapist training or client utilization than are many other modalities, such as Jungian or Freudian psychoanalysis. This chapter challenges us as therapists to examine how we can expand the availability of and accessibility to sandplay as a therapeutic option.

Chapter 4 focuses on transference and countertransference—cotransference—from an ethical, rather than a clinical perspective. Loue notes here that a failure to address such issues competently may potentially subject a client to harm, contravening the ethical principles of beneficence and nonmaleficence.

The growing interest in sandplay research suggests the need to examine the ethical issues that may accompany such investigations. Loue addresses in Chap. 5 the general ethical principles that govern research involving human participants, with reference to the ethical codes of various mental health professions.

Loue and Parkinson highlight not only the ethical issues, but also the legal issues that are often associated with dual relationships and conflict of interest in general, and in the context of sandplay therapy specifically. Parkinson and Loue focus in Chap. 7 on special ethical consideration in sandplay therapy practice. The current emphasis on evidence-based practice would have us believe that the absence of rigorous systematic examination of sandplay therapy's efficacy suggests its ineffectiveness. Such a conclusion would be both uninformed and ill-founded. Nevertheless, because sandplay has not been subject to rigorous, systematic evaluation, respect for clients, through the informed consent process, suggests that they be made aware of any associated risks of the practice. Various other issues common to other therapeutic modalities are examined here in the context of sandplay therapy practice: abandonment, dual relationships, and conflict of interest.

We draw in each of these chapters from well-enunciated ethical principles of clinical practice and research. We have integrated to the extent possible relevant portions of professional ethics codes governing mental health professionals from an array of English-speaking countries—Australia, Canada, the UK, the USA, and New Zealand—to maximize the relevance of the text to sandplay therapists worldwide, whether they are licensed as psychologists, marriage and family therapists, counselors, or social workers.

A danger in editing such a book as this is that it will be viewed by those within the profession and the professional societies as an indictment or accusation that something is not quite right. Such is not my intent, nor is it the intent of the authors who have contributed their work to this space. I would wish instead that the thoughts and opinions contained here will prompt each of us to examine further who we are as individuals, as therapists, and as professionals within our organizations; how we can grow in each of these capacities; and how we can move the field of sandplay therapy forward into the future.

Cleveland, Ohio   Sana Loue
September 2014

References

Weinrib, E.L. (1983). Images of the self: The sandplay therapy process. Boston, MA: Sigo Press.

Acknowledgments

It has been and continues to be a gift and a privilege to engage with clients and colleagues through sandplay therapy. I cannot acknowledge by name the many clients from whom I have learned and through whose sharing I have been transformed, but I thank them here anonymously. Their courage in sharing their experiences and their journeys prepares me to assist not only them, but those clients who come after them.

I am indebted to all of the many colleagues in sandplay therapy from whom I have learned over the years, through informal conversations, conferences, consultation, trainings, writings, and yes, my own personal process. I want to thank, in particular, Jill Kaplan, who has been serving as my advisor through my own certification process and has unstintingly shared her wisdom and her insights, and challenged me to go beyond my own thinking. I marvel at Lucia Chambers' intuitive skills as a therapist, her ability to skillfully plummet to the depths of meaning reflected in a client's sandplay experience. Lucia embodies a way of being in the world that conveys, at the deepest level, the free and protected space of sandplay. Judy Zappacosta has inspired me through her writings, precise in meaning and poetic in form that continually capture the essence of what sandplay is. Regina Driscoll marvelously integrates technology and sandplay and it is from her that I have learned some tidbits about Celtic symbology. Others, too numerous to name, have shared portions of their journeys through life and/or towards certification with me, allowing me to see the many possibilities.

Contents

Contributors

Carina Conradie New Zealand, New Zealand

Roxie Hanes New Zealand, New Zealand

Sana Loue Case Western Reserve University, Cleveland, OH, USA

School of Medicine, Case Western Reserve University, Cleveland, OH, USA

Jean Parkinson Auckland, New Zealand

About the Authors

Carina Conradie is a Registered Clinical Psychologist in New Zealand who emigrated 10 years ago from South Africa where she originally registered as a psychologist specializing with children in 1995. Over the years, she has always had a private practice. In New Zealand she has been working part-time for the Ministry of Education. a charitable organisation and is currently at an Independent School with Roxie Hanes one day a week. Her clientele has always been mainly young people ranging from three years to adolescents. More recently, she has also had some adult clients. Sandplay therapy is the main modality in which she works.

Carina completed her first training in Sandplay therapy in South Africa in 2003. Over the last six years, she has committed herself to working towards becoming a certified member of the International Society of Sandplay Therapy (ISST). This involves going overseas to be trained by registered ISST teachers as well as a two-week intensive training course in Switzerland. The lack of funds and having registered members in New Zealand delay the completion of this goal. Carina has been getting external SKYPE supervision in Sandplay Therapy by a registered ISST teacher over the last three years.

Roxie Hanes is a Registered Psychologist in New Zealand within the Scope of Practice of Community Psychology. Her practitioner journey started in 1978 and she completed her registration as a psychologist in 2004. She has worked with students with learning and behavioural difficulties both in New Zealand and internationally. She has also had several years of working with adults with mental health difficulties in various organisations and settings. Roxie currently works as a psychologist and is the manager of the Psychology and Counselling Service at an Independent School for students from years 7-13. Many of the students have high, complex needs such as anxiety or depression, engaging in self-harm and experiencing suicidal ideation.

Roxie was introduced to sandplay therapy by her colleague Carina Conradie, her co-author in this volume. Roxie attended a one-day introduction course and a 2-day Level 1 training. She has had supervision with Carina specifically relating to Sandplay and Sandtray therapy. Roxie explains that her expertise in this field is much less than Carina's but she feels that she practices safely within her scope of practice and with supervision and guidance from Carina.

Sana Loue is a licensed independent social worker (LISW) in Ohio, USA. She holds graduate degrees in social work (MSSA), law (JD), epidemiology (PhD), medical anthropology (PhD), public health (MPH), education (MA), and theology (MA). Dr. Loue is a faculty member at Case Western Reserve University School of Medicine, with appointments in the departments of Bioethics, Psychiatry, and Epidemiology and Biostatistics and the Center for Global Health. Her clients range in age from their mid-teens to their mid-fifties, many of whom have diagnoses of depression, anxiety disorders, schizophrenia, and sexual trauma. The majority of her clients self-identify as gay, lesbian, or transgender. She has authored manuscripts and books in a wide variety of fields; several of her manuscripts have appeared in the Journal of Sandplay Therapy.

Jean Parkinson is a registered art psychotherapist (AThR, ANZATA), a sandplay practitioner, and an ISST candidate. She is engaged in private practice in Auckland, New Zealand with children, teens and adults and as a Clinical Supervisor. Special interests are trauma, attachment, relationships and neuropsychology. Her counseling training is in Choice Theory/Reality Therapy/Quality Management and Narrative Therapy and Jean retains her Teachers' registration. She was on the ANZATA committee for 5 years, involved with Supervision and Ethics subcommittees.

Chapter 1
Are We There Yet? Ethical Issues Associated with Self-Identifying as a Sandplay Therapist

Carina Conradie and Roxie Hanes

1.1 Introduction

'Are we there yet?' How many of us have asked this question on a long journey? One of our mother's usual response of 'another 6 h' often left the kids disappointed and unmotivated. The journey of becoming a registered or licensed sandplay therapist is also long and taxing and not many professionals reach the point of registration and/or certification to their national or international body. The journey may begin and end for some with a short 3-h introduction course to sand tray—or it may involve an epic journey involving years of travelling to get training and supervision by accredited international teachers in sandplay therapy. For us in New Zealand, training opportunities are limited, resulting in professional practitioners having very different journeys and final destinations. In this chapter, we discuss the ethical dilemmas we have encountered working in agencies where not everyone is on the same journey. We also explore the implications that the varied therapist journeys may have for the uninformed client; our own professional bodies; our co-workers, as well as management or supervision, that is being managed and/or supervised (or managing and supervising) by someone who is on a completely different path than our own with respect to their expertise, training and supervision.

1.2 Sandplay and Its Practice

For the purpose of this chapter, *Sandplay Therapy* is defined by the International Society of Sandplay Therapy (ISST) as a therapeutic method developed by Dora Kalff and based on the psychological principles of C. G. Jung. Sandplay therapy is a creative form of therapy in using imagination, 'a concentrated extract of the life

C. Conradie (✉) · R. Hanes
New Zealand, New Zealand
e-mail: iconradie@xtra.co.nz

© Springer International Publishing Switzerland 2015
S. Loue (ed.), *Ethical Issues in Sandplay Therapy Practice and Research,*
SpringerBriefs in Social Work, DOI 10.1007/978-3-319-14118-3_1

forces both physical and psychic' (International Society of Sandplay Therapy 2006, Art. 2). It is characterized by the use of sand, water and miniatures in the creation of images within the 'free and protected space' of the therapeutic relationship and the sand tray. A series of sandplay images portrayed in the sand tray creates an ongoing dialogue between the conscious and the unconscious aspects of the client's psyche, which activates a healing process and the development of personality (ISST Statutes 2006, Art. 2). It is a nonverbal understanding of images that helps the therapist stay attuned to what is going on; for this reason, verbal interpretations are usually delayed. In sandplay, dual processes take place—firstly, the analytical process and, secondly, the deliberate regression into the preconscious, preverbal matriarchal level of the psyche (Bradway and McCoard 1997). For the purpose of this chapter, *sand tray* refers to other therapeutic uses of sand, water and miniatures, for example where the client is instructed to make a tray, e.g. a dream or to choose miniatures that would represent certain emotions.

The ISST is an umbrella society operating through a number of national societies that exist throughout the world, with members who are therapists trained in sandplay therapy and who meet the guidelines defined by the society. These guidelines were formulated in order to protect the high professional standard of the work. The ISST guidelines focus on both the training in sandplay therapy and the requirements for becoming a teaching member of ISST. The organisation also serves as an international forum for the exchange of professional experience with sandplay therapy.

Article 5 of ISST statutes (ISST 2006) specifies that there are three categories of ISST members:

1. The national societies (legal persons): National societies are sandplay associations that are accepted as the national societies of ISST. The national societies are members of ISST; their ISST-certified members are indirect members of ISST. The norm is one national society per nation.
2. Individual, direct members (natural persons): An Individual member is a sandplay therapist who is accepted into the ISST and is not a member of a national society. This status is valid for:

   a. Sandplay therapists who practice in a geographical area where there are not enough (min. 5) ISST-certified sandplay therapists for the establishment of the national society
   b. Particular circumstances which allow for exceptions subject to the decision of the Board

3. Honorary members: These are, first and foremost, the founding members.

The Sandplay Therapists of America (STA) is one of these national societies. The STA's *Handbook of Practitioner Member Requirements and Procedures* (2013) specifies that therapists seeking certification as a full member have undergone a personal process of a minimum of 40 sessions; completed a comprehensive programme of study (minimum of 120 h) covering introduction to sandplay therapy, Jungian theory and symbolism and clinical sandplay practice; received a minimum of 80 h of consultation (individual and group) with an ISST member and authored two preliminary papers and a final case study that have been reviewed and found to be acceptable by the requisite number of fully certified sandplay therapists (Handbook of Certified, Teaching and Practitioner Member Requirements and Procedures for Sandplay Therapists of America).

Sandplay Therapists Association of Australia and New Zealand (STANZA) is currently a working group of sandplay therapists working towards ISST accreditation. At this stage, STANZA offers a collegial forum and a venue for professional development for practitioners who are working towards registration as ISST members.

Considering the guidelines defined by the various national and international societies, it is clear that the process of becoming a professional registered sandplay therapist is extensive and arduous, requiring significant training by registered teaching professionals, hours of consultation or supervision (individually and in group situations with a registered teaching member) as well as a personal process by a professional—the long epic journey!

We are two registered psychologists with very different levels of expertise in general psychology training and specifically in sandplay and sand tray therapy. Roxie is a registered community psychologist and full-time manager of a psychology and counselling service for a big organisation working with clients from 10 years of age to adults. Carina is a registered clinical psychologist working part-time in the same organisation and also in a private practice. In the organisation, we also have co-workers with counselling backgrounds. It is interesting to note in our situation that the manager is less experienced in sandplay therapy; therefore, we can really speak from a place of honesty and integrity about the complexities of this dilemma. We also speak from experience of working in other multidisciplinary organisations.

The following are headings and considerations of the complexities of practical applications and experiences that we have encountered in a New Zealand context and in our day-to-day work.

1.2.1   Levels and Limitations of Training and Experience in Sandplay and Sand Tray

Sandplay therapy is a therapeutic method developed by Dora Kalff and based on Jungian psychological principles. In New Zealand, most universities offer clinical programmes to become a registered psychologist. Most of these programmes have an extensive theoretical foundation in related mental health and in developmental, cognitive and behavioural theories. Sandplay therapy and Jungian psychology are not offered as a specific modality in New Zealand universities. There are various counselling programmes that range from short course to certificate, diploma, degree and possibly a postgraduate degree; anyone can basically call themselves a counsellor. Consequently, the depth of knowledge gained in training varies immensely. Some institutions that offer creative therapies offer students 'taster' courses in sandplay, sand tray and other associated symbol work.

Sandplay therapy and Jungian psychology are not offered as specific modalities in New Zealand universities. In New Zealand, there is no formal training in sandplay and/or sand tray therapy. Therefore, every professional who is interested in this modality needs to go to extraordinary lengths to receive training and appropriate supervision. It really comes down to the professional's own integrity as to whether a professional therapist feels adequate to practice and to reduce the risk of harm for the client. As a modality, it is not well known and/or valued—for us it has become a lonely but satisfying journey.

In our experience, some people who have undergone these trainings (maybe 2 h of teaching) feel confident to do sandplay therapy and call themselves sandplay therapists. The uninformed client would know no better and would be unaware of the potential risk associated with engaging in sandplay therapy with individuals having so little experience with this modality.

In New Zealand, health practitioners are registered and governed by the Health Practitioner's Competence Assurance (HPCA) Act 2003. The principal purpose of this Act is to protect the health and safety of members of the public by providing mechanisms to ensure that health practitioners (including psychologists) are competent and fit to practice their professions. Practically, this means that as a psychologist, one must provide evidence in a yearly continuing competency plan that demonstrates competence or develops competence in the modalities that one is utilizing, for the purpose of this discussion, sandplay therapy. Immediately as psychologists, we are limited as there are not many professional development opportunities in New Zealand and very few agencies will support training financially. It becomes the practitioner's own commitment to sandplay therapy that will drive this development.

The HPCA Act further states:

> No health practitioner may perform a health service that forms part of a scope of practice of the profession in respect of which he or she is registered unless he or she—(a) is permitted to perform that service by his or her scope of practice; and (b) performs that service in accordance with any conditions stated in his or her scope of practice.

The Psychologists' Registration Board is not familiar with the modality of sandplay therapy and therefore must rely on the practitioner's own integrity, conscience and honesty in recording his or her professional development in sandplay therapy in his or her portfolio and gaining adequate supervision for it (See New Zealand Psychologist Board 2011). This deepens the ethical dilemmas and risks that psychologists practising sandplay therapy may have. It is a high trust model. It seems that secondary modalities such as sandplay and sand tray most often sit outside of the scopes of practice that are subject to quality assurance. Sandplay therapy is 'off the beaten track' in New Zealand. Consequently, members of the general public cannot assume that registration or non-registration as a health practitioner affords any greater quality assurance with regard to the modality of sandplay and sand tray therapy.

1.2.2 Supervision

Supervision is a required competency for all registered psychologists and the counsellors who wish to have membership to a professional body in New Zealand.

Supervision is a contractual process involving a supervisor and supervisee meeting on a regular basis to enhance psychology- and/or counselling-based work and/or professional functioning. The purpose of the supervision relationship may vary and can be peer, mentoring, training and or evaluative (Code of Ethics for Psychologists working in Aotearoa/New Zealand 2012). This can be done internally, meaning that the organisation provides a supervisor who also works for the same organisation, or externally, meaning that the practitioner has supervision by someone outside the organisation. In some cases, practitioners can choose a supervisor and in other cases not. Possible reasons for having an external supervisor are:

- No one within the organisation has similar training
- Budgetary concerns
- Requirements of the professional organisational body
- The philosophy of employer
- Employment contract

An ethical dilemma occurs when a professional practices sandplay therapy and sand tray therapy as one of a range of counselling techniques, with or without training, and his or her supervisor has limited or no knowledge or experience in this modality. The supervisor may even inform the supervisee that he or she feels competent to supervise this modality. The supervisor can easily be in a position of power. The client who is doing the sandplay therapy, as well as the inexperienced supervisee, may be at risk.

Risks to the client may include premature or inappropriate interpretations of sandplay work and sharing this information with the client, which may change and/or influence the direction of therapy and potential healing. A therapist with an inexperienced supervisor in sandplay therapy cannot get appropriate guidance on how to move forward with client work. This may consequently hinder the client's process of healing. Central to Dora Kalff's sandplay therapy is the concept of the free and protected space, which has both physical and psychological dimensions (Weinrib 2004, p. 29). A practitioner who has an untrained supervisor and is therefore unable to hold a 'free and protected space' for the supervisee can indirectly harm the client who is dependent on the therapist's expertise and supervision. Inexperienced therapists may erroneously assume or believe that they are receiving appropriate and professional supervision in the progression of their own development in sandplay therapy.

A risk for the supervisor may be 'delusion and/or denial' that he or she is competent to be supervising someone in a modality in which he or she lacks adequate skill and understanding. If the supervisor is not trained in sandplay therapy, he or she most probably lacks the necessary resources for sandplay, such as sand, water and figurines. Consequently, the therapist seeking supervision will not have an opportunity to experience sandplay therapy and will have a severely limited understanding of what sandplay clients actually experience during their process. One can analogize this situation to that of a travel agent who has never climbed a significantly high mountain and who advises an intrepid traveller that the only requirements for climbing Mount Everest are enthusiasm, a warm coat and a guidebook!

1.2.3   Level of Practitioner Confidence

As sandplay therapy requires considerable training and supervision in addition to a strong foundation of psychological knowledge, competence should match confidence. We have found that this is not always the case. On the one hand, a supremely confident practitioner with limited training and skills may practice outside of his or her scope of practice and the employing organisation may support this due to its lack of knowledge and its focus on monetary concerns. An overconfident practitioner may also be unable and or unwilling to reflect on the sandplay process of the client and how the practitioner's inexperience and lack of understanding may not provide the requisite free and protected space. On the other hand, a competent practitioner may lack confidence to practice sandplay therapy and may therefore limit the possibility of the healing sandplay offers. In both of these situations, practitioner's engagement in his or her own sandplay process would solve this mismatch of competence and confidence. However, in our situation, there is limited access to experienced sandplay therapists.

In New Zealand, the HPCA Act 2003 provides legislation to protect the public by ensuring health professionals including psychologists practice within their scope of practice and their level of competence. The HPCA Act states the following:

a. For the determination for each health practitioner's scope of practice within which he or she is competent to practise.
b. For systems to ensure that no health practitioner practises in that capacity outside his or her scope of practice.

The HPCA Act requires competence over mere confidence. The dilemma, however, is that not all practitioners doing sandplay therapy are identified as health professionals; for example, counsellors are not registered as health practitioners and therefore are not held accountable under the HPCA Act.

1.2.4   Limited Opportunities for Practitioners to do Their Own Sandplay Work

Sandplay within a free and protected space offers the psyche on a deep unconscious level—an autonomous opportunity to heal itself. Through the sand, water and figurines, a connection can be made between the outer and inner world. The process enables the constellation and activation of the self, the subsequent wounding of the ego and the recovery of the inner child (Weinrib 2004, p. 2).

How can we expect our client to experience this deep level of healing if we have very little experience of the sandplay process ourselves as practitioners? In our experience, due to the nature of training in New Zealand, reflection is a commonly used term; however, the skill required to reflect one's own process is not often articulated. If the practitioner has not undergone his or her sandplay process (which is in most situations the reality), how can he or she truly understand what the client is experiencing? A supervised experience in creative therapies may be helpful. As mentioned before, having a supervisor who cannot offer this modality in supervision will limit the supervisee's experience in sandplay therapy. *Are we there yet* almost becomes an impossibility within the New Zealand context.

1.2.5 Note Taking

In most agencies, some degree of recordkeeping of notes is required for the purposes of accountability, complaints and a history of the service provided. These notes are confidential but are usually accessible for all practitioners within an organisation. In our experience, we have found inexperienced colleagues reading notes of experienced practitioners as a way of learning and developing their own skills. However, some have then made similar assumptions about symbolism and/or process across one client to another, a practice that is neither appropriate nor safe. Again, sufficient and appropriate supervision or peer mentoring may prevent this and offer the therapist wonderful learning opportunities. In our experience working in a multidisciplinary agency, we do inform our clients that notes are confidential but may be read by colleagues for the purpose of continuity of care.

Expressed in a different way, a problem we have found is that counsellors often less experienced in sandplay therapy and sand tray work, who want to grow and develop in this modality, can use notes that other colleagues have written to appear more competent and confident than they actually are. The client, however, may not be informed that this is occurring and, consequently, issues relating to a breach of privacy and confidentiality arise. In this scenario, there is no guidance and therefore it is debatable whether it is for the 'purpose of training.' No one is guiding the training and the external supervisor does not know that this is happening.

It is not possible to monitor this and one has to rely on honesty and integrity. Again, if the agency provides peer-mentoring opportunities where learning can take place in a transparent environment, this dilemma may be lessened. If peer mentoring is seen as a conflict of interest by one of the team members, it may further exacerbate the problem.

From a line management perspective, questioning about the observations or analysis written in the notes can be taken as criticism rather than learning. If the external supervisor does not read the notes, there is no knowledge of this potential risk.

1.2.6 Multiple Roles

In our experience working in multidisciplinary agencies, there are often multiple roles, e.g. social workers, counsellors, psychotherapists and psychologists, and then there are various levels of skill within a specific modality, e.g. sandplay. For example, a social worker may have completed the most training in sandplay but the psychologist, who has limited training in sandplay, oversees the work of the social worker and might be his or her line manager. Accordingly, there may be multiple hierarchies of expertise; how this is managed becomes extremely important. The person who has greater expertise in sandplay should remain the expert in sandplay, while the line manager remains the person who has overall responsibility for the individual's work and the work of the team. In our experience, it can work very well when there is open transparent awareness, but it can also become problematic if there is not an understanding of the need for sandplay expertise to be seen as separate from line management. Clear boundaries between the different roles and levels of responsibility can deal with this effectively.

Another dilemma that may arise is when a practitioner becomes rigid in his or her view of 'conflict of interest' and declines supervision from an experienced sandplay therapist/colleague. The integrity and flexibility of the practitioner may determine the safe management of risk for the client and potential growth for themselves.

1.3 Summary

For Carina and Roxie, our journey will end at different destinations. In writing this chapter, we have concluded that the 'Are we there yet?' is not as important as providing a safe and best practice for our clients and the uninformed public. We are committed to continue our journeys at different paces and with different destinations and acknowledge that we continually rely on each other as fellow travellers. Recognising the stages of the journey as well as the obstacles we have en route is important. The HPCA Act 2003 in New Zealand has not alleviated the ethical dilemmas we have discussed above. Our general New Zealand travel insurance does not cover this journey—we need to cover ourselves additionally!

This metaphorical insurance may provide us with a middle road—relying on our own integrity, commitment and professionalism. *Are we there yet* may have to change to *Are we travelling safely* and respectfully with our clients and colleagues.

1.4 Appendix

In New Zealand, there is a government act that promotes the safety of clients. The act is entitled the Health Practitioners Competence Assurance (HPCA) Act (2003). Relevant sections of the Act are provided below.

1.4.1 Sections of the HPCA Act

3 Purpose of Act
1. The principal purpose of this Act is to protect the health and safety of members of the public by providing for mechanisms to ensure that health practitioners are competent and fit to practice their professions.

2. This Act seeks to attain its principal purpose by providing, among other things:

(a). For a consistent accountability regime for all health professions
(b). For the determination for each health practitioner of the scope of practice within which he or she is competent to practise
(c). For systems to ensure that no health practitioner practises in that capacity outside his or her scope of practice
(d). For power to restrict specified activities to particular classes of health practitioner to protect members of the public from the risk of serious or permanent harm
(e). For certain protections for health practitioners who take part in protected quality assurance activities
(f). For additional health professions to become subject to this Act

8 Health Practitioners Must Not Practise Outside of Scope

1. Every health practitioner who practises the profession in respect of which he or she is registered must have a current practising certificate issued by the responsible authority.
2. No health practitioner may perform a health service that forms part of a scope of practice of the profession in respect of which he or she is registered unless he or she:

(a). Is permitted to perform that service by his or her scope of practice
(b). Performs that service in accordance with any conditions stated in his or her scope of practice

3. Nothing in subsection (1) or subsection (2) applies to a health practitioner who performs health services:

(a). In an emergency
(b). As part of a course of training or instruction
(c). In the course of an examination, assessment or competence review required or ordered by the responsible authority

References

Bradway, K., & McCoard, B. (1997). *Sandplay-silent worksop of the psyche*. New York: Routledge.
Health Practitioners Competence Act 2003., Public Act No 48, Date of assent 18 September 2003. http://www.legislation.govt.nz/act/public/2003/0048/latest/DLM203312.html. Accessed 3 Feb 2015.
International Society of Sandplay Therapy. (2006). Statutes. http://www.isst-society.com/homeng.php?site=statutes. Accessed 1 Feb 2014.
New Zealand Psychologist Board. (2011). *Core competencies for the practice of psychology in New Zealand*. Wellington: New Zealand Psychologist Board Registration Committee. http://www.psychologistsboard.org.nz/cms_show_download.php?id=†41. Accessed 11 Feb 2014.

New Zealand Psychologist Board. (2012). *Code of ethics for psychologists in Aotearoa/New Zealand*. Wellington: New Zealand Psychologist Board.

Sandplay Therapists of America. (2013). Handbook of certified, teaching and practitioner member requirements and procedures for Sandplay Therapists of America (ISST). http://www.sandplay.org/pdf/STA_Handbook.pdf. Accessed 14 Feb 2013.

Weinrib, E. L. (2004). *Images of the self: The sandplay therapy process*. Cloverdale: Temenos.

Chapter 2
Ethical Issues in Sandplay Cyber-Supervision

Jean Parkinson and Sana Loue

2.1 Introduction

It has been estimated that 34.3 % of individuals worldwide and 77 % of adult Americans now utilize electronic mechanisms for communication (Internet World Stats 2013; Pew Research Center 2014). Health care providers, and mental health care providers specifically, now utilize various forms of electronic media to communicate with their colleagues. These communications between colleagues may be for the purpose of seeking advice regarding a particular situation, a particular client, or as part of an ongoing supervisory–consultative relationship.

The use of electronic means for professional supervision-consultation may be particularly important for sandplay therapists. Therapists utilizing sandplay may not accrue hours towards certification in this modality without supervision from sandplay therapists who are certified as teaching members by their national (Sandplay Therapists of America 2012) or international organization International Society for Sandplay Therapy (ISST). Yet, in many geographic areas, there are no certified sandplay therapists who can provide such consultation. As a result, the sandplay therapist seeking certification hours or seeking consultation in an effort to provide competent mental health care to his or her clients must often travel great distances if face-to-face consultation is to occur. In many circumstances, this is not feasible due to the distance that must be traveled and the costs and time associated with such travel. As an example, one author of this chapter (JP) practices in New Zealand but must seek consultation from an ISST-certified teacher outside of Australia and New Zealand to meet ISST certification requirements because there are currently no

J. Parkinson (✉)
Auckland, New Zealand

S. Loue
Case Western Reserve University, Cleveland, OH, USA
e-mail: sana.loue@case.edu

© Springer International Publishing Switzerland 2015
S. Loue (ed.), *Ethical Issues in Sandplay Therapy Practice and Research,*
SpringerBriefs in Social Work, DOI 10.1007/978-3-319-14118-3_2

ISST-certified teachers in Australasia. The second author (SL) has obtained consultation services from more experienced practitioners in Minnesota and in California due to the nonexistence of certified sandplay practitioners nearby.

This chapter discusses the potential benefits that can be derived from cyber-supervision and explores the ethical issues associated with the use of electronic means for the purpose of supervision. The chapter concludes with a summary of recommended practices for the supervisee and the supervisor.

2.2 Defining Cyber-Supervision

Clinical supervision has been described as:

> a collaborative process that occurs between a more experienced and skilled supervisor and a novice or apprentice trainee—the supervisee—who seeks to develop the competencies necessary for successful clinical practice. (Barnett 2011, p. 105)

It is a process that involves "observation, evaluation, feedback, the facilitation of knowledge and skills by instruction, modeling, and mutual problem solving" (Falender and Shafranske 2004, p. 3). The process of supervision is deemed critical to the training of mental health professionals (Barnett et al. 2007, p. 273; Romans et al. 1995, p. 407). In this chapter, we use the terms supervision and consultation interchangeably to denote the consultative relationship between sandplay therapists for the benefit of the client.

However, that supervision or consultation does not occur only between those who are formally engaged in training programs and those who are more experienced in the mental health field, such as between graduate students and their fieldwork supervisors. Consultation or supervision occurs—and should occur—on a regular basis between a mental health care provider and a colleague within the context of a formalized consultation–supervision relationship. (We recognize, however, that depending upon one's legal jurisdiction, the legal liability of a supervisor may differ significantly from that of someone providing consultation services).

Supervision of sandplay therapy presents a third dynamic—a visual image of unconscious processes and creative imagination:

> The unfolding of a series of sand creations also allows us to view the vastness and complexity of the unconscious. Through study of sand pictures, we are able to identify the development of the relationship between the ego and the Self, the journey toward individuation, bridging and integration of unresolved issues (i.e., tension of opposites) emergence of new creative energies, and movement towards wholeness.... Also, when the supervisor highlights the supervisee's unique emotional and intuitive responses then the supervisee's own approach emerges. In this safe environment, therapists' individual gifts and talents are validated and allowed to flourish. (Friedman and Mitchell 2008, p. 4)

Indeed, it appears that inadequate supervision may lead to lowered job satisfaction and burnout (Jerrell 1983).

Traditionally, the supervision or consultation process has occurred in a face-to-face relationship between the supervising mental health care provider and the supervisee. As technology has developed and professionals' comfort level with it has increased, technological means, such as telephone and fax, have been utilized to augment this relationship (VandenBos and Williams 2000). Materials associated with a client's therapy that required viewing, such as drawings done in the context of art therapy or a sand tray made during a sandplay therapy session, were often photographed and copies sent via US mail from the consulting provider to the supervisor and were returned to the provider by mail following the telephone consultation session. Videotapes of sessions could also be sent to the supervising provider via US mail and returned to the supervisee after telephone supervision sessions had concluded (Wetchler et al. 1993). More recently, mental health care providers, including sandplay therapists, have been using Internet-based mechanisms for supervision. VandenBos and Williams reported in 2000 that 2 % of the 596 supervising psychologists participating in their study had utilized the Internet or satellite technology for supervision purposes.

To the best of these authors' knowledge, the term "cyber-supervision" has not been previously employed. We define it to encompass the use of Internet for the purpose of supervision or consultation by one mental health care provider to another. These communications can be:

> to obtain or provide consultation to or from a colleagues; for the provision of clinical supervision across distances; and to offer psychotherapy and supervision training in situations where in-person training is not feasible as well as when the use of various technologies may enhance the quality and effectiveness of the in-person training provided. (Barnett 2011, p. 103)

These Internet communications may be effectuated via e-mail, chat rooms and instant messaging, videoconferencing programs, and televideoconferencing systems. Mechanisms such as e-mail are asynchronous in that, while they allow the instant delivery of a message, the response to such messages may be time-delayed at the discretion of the recipient. In contrast, mechanisms such as chat rooms and instant messaging are synchronous, permitting users to respond to each other in real time. Televideoconferencing, such as through Skype or Oovoo, can allow the supervisor to observe a session between the therapist and the client in real time; face-to-face consultation may occur immediately following the client's departure from the session or through messaging during the actual session (Neukrug 1991; Smith et al. 1998). Documents to be viewed as part of the consultation–supervision, such as photos of sand trays, are often transmitted via e-mail attachment or by sharing a file in Dropbox, Google Groups, or the iCloud.

2.3   The Promise of Cyber-Supervision

Cyber-supervision may be particularly helpful, and even necessary, in a variety of situations. Individuals who are practicing in geographic areas that are relatively isolated or in which there are no other providers trained in a specific therapeutic modality, such as sandplay, may need to look far afield to identify a colleague who can provide competent supervision. Mental health providers located in rural areas, for example,

> experience pressures, both from within themselves and from their communities, to try to be everything to everyone in order to meet what sometimes seem like overwhelming needs. Some quickly educate themselves by using Internet resources and other methods of distance learning or by reading books and journal articles in an attempt to learn along the way. (Schank 1998, p. 275)

Absent access to and utilization of competent supervision, mental health providers may unwittingly stray outside the scope of their practice, thereby increasing the potential harm to the client. (For a discussion of the ethical issues associated with practicing or supervising outside of one's practice scope, see Conradie and Hanes, this volume.) Sandplay therapists may take on a client who is from a very different culture or one whose primary language is not the same as that of the therapist. Supervision from a provider with greater familiarity with the client's culture or an understanding of the client's primary language may be critical to the provision of competent care (cf. Kanz 2001).

The specific benefits said to be associated with the use of cyber-supervision vary, to some degree, depending on the specific modality to be used. E-mail, chat rooms, and instant messaging, unlike televideoconferencing, do not require a face-to-face contact between the supervisor and the supervisee. These mechanisms may, therefore, provide either the supervisor or the supervisee or both with a sense of psychological safety (cf. Zuboff 1988). It has been suggested that "reading and writing through e-mail may involve a unique personal mechanism that facilitates self-disclosure, ventilation, and externalization of problems and conflicts and that promotes self-awareness" (Barak 1999, p. 237). Additional benefits attributed to the use of e-mail include the individual's disinhibition (Joinson 1998) so that supervisees may be more willing to disclose personal feelings and issues (Stebnicki and Glover 2001), an increase in supervisee reflectivity (Clingerman and Bernard 2004), and an increased sense of support among supervisees (Stebnicki and Glover 2001).

2.4   Ethical Issues

2.4.1   Confidentiality, Privacy, and Client Informed Consent

Although cyber-supervision potentially offers some benefits and, in some circumstances, may be the only feasible means of obtaining and providing supervision, confidentiality and privacy constitute major issues. We use the term confidentiality here to refer to the information collected or compiled and documented about an individual, regardless of the form of that documentation. Privacy, in contrast, refers to the individual himself or herself. It is clear that confidentiality and privacy are of utmost importance and are ethically required of sandplay therapists. The Code of Ethics of the ISST (n.d., para. B.1) provides: "ISST members and candidates respect client rights to privacy and do not share confidential information without client consent or without sound legal or ethical justification."

Records relating to clients, whether they were insurance forms, clinical note, or clinical reports, once existed in some form of hard copy only. These were generally safeguarded by limiting access to the records, storing the information in locked cabinets in locked offices, and minimizing their inadvertent transmission to others through the use of mail and fax. Such records were generally stored for a predetermined period of time following the cessation of the therapeutic relationship and then destroyed by burning or shredding them. Photographs of client work were often taken with a Polaroid camera; no duplicates existed and these, too, were often similarly destroyed.

More recently, many therapists rely on computers and iPads to record their clinical notes; some may use the note-taking feature on their iPhones. Many sandplay therapists use their iPads and/or iPhones to take and store images of client sand trays. Backup copies may exist on flash drives or in the Cloud. While these storage mechanisms may facilitate the transmission of material between the supervising therapist and the supervisee, they raise vexing concerns related to the therapist's ability to safeguard confidentiality. The difficulties associated with protecting electronic transmissions of client information do not relieve the sandplay therapist of the ethical responsibility to maintain client confidentiality and privacy. As stated by the ISST's Code of Ethics (n.d., para. B.6.):

> ISST members and candidates take precautions to ensure the confidentiality of information transmitted electronically, including but not limited to electronic mail, voicemail, answering machines, facsimile machines, and websites.

One basic mechanism that can be used to safeguard data housed on computers, iPhones, iPads, and flash drives is the use of a password. Care must be taken, however, in the formulation of a password. Many people use the same password for multiple functions in order to reduce the risk of forgetting it. However, this practice increases the risk that client records could be compromised if the password were to be inadvertently disclosed or deliberately hacked. Passwords that are weak, with too little complexity, may also result in an increased likelihood of discovery by unauthorized persons. Hackers can penetrate an individual's Cloud account or e-mail account, thereby becoming privy to the sensitive details of the client's life.

E-mails present additional challenges to the maintenance of confidentiality. Many agencies maintain policies and procedures that permit designated individuals within the organization to access and read all e-mail correspondences generated from or sent to organization-based e-mail addresses or that are sent to or from equipment owned, purchased, or leased by the organization; these individuals are likely not involved with the client's care. As an example, the faculty handbook of the University of Florida provided, "All electronic mail records are public records and are stored in memory by the Northeast Regional Data Center" (University of Florida Office of Academic Affairs 1993). Case Western Reserve University (n.d.) advises its faculty, employees, and students, "There should be no expectation of an inherent right to privacy—such rights cannot be guaranteed within the myriad IT uses at Case." The nonuse of screen savers by the therapist and/or the supervising therapist may allow others in their offices to view client materials to which they should not be privy. The use of programs that automatically complete a recipient's e-mail address may inadvertently lead to the misdirection of the e-mail. And, there is the ever-present possibility that an e-mail account will be hacked, an event that is occurring with increasing frequency. As one writer noted, "All e-mail messages can be read by people other than their intended recipients, so one must assume that they will be read, even if it isn't actually the case." (Glossbrenner 1990)

Accordingly, it is critical that, if cyber-supervision is to be utilized, additional mechanisms be implemented to safeguard client confidentiality. In addition to complex passwords, supervisors and supervisees should use passworded screensavers when away from their desks to prevent others in their offices from viewing confidential materials. Encryption software can be utilized to reduce the likelihood that a hacker will be able to read the content of the transmission. Virtual private networks (VPN) can be utilized to reduce the likelihood that transmissions effectuated over public networks can be read by unauthorized persons. Firewalls can reduce the risk of security breaches.

Sandplay therapists who work in a group practice or for an agency or hospital may have only some control, or no control, over the measures that are utilized to protect data. The implementation of inadequate measures by an institution potentially subjects all data to the risk of unauthorized disclosure (Schultz 2012). In such situations, the therapist remains ethically—and possibly legally—responsible for any unauthorized disclosures that may occur, but may have had limited ability to prevent their occurrence.

The portability of the devices on which client information is stored raises yet additional issues. Laptop computers, electronic tablets, smartphones, and flash drives are all subject to inadvertent loss and potential theft. Absent adequate protective measures, unscrupulous finders or thieves may utilize the information they find to their own advantage and to the detriment of the client. Many mobile phone and tablet manufacturers have added features to newer electronic devices that allow their owners to lock the devices remotely when they are lost or stolen. We highly recommend that both supervisors and supervisees utilize these mechanisms whenever available, in addition to implementing the other security measures noted.

Client privacy is also subject to heightened risk of violation as the result of electronic transmissions. The real-time transmission of a sandplay session from its occurrence in the therapist's office to the observing supervisor may be intercepted by a hacker, who may then rebroadcast the content and/or use it to his or her own advantage. Once any material is sent through cyberspace, both the therapist and the client lose control over its further dissemination (Kanz 2001).

Case Example 1

A sandplay therapist engages a more senior, experienced sandplay therapist in another country for consultation on a regular basis. Some form of cyber-supervision is necessary because there are no certified, experienced sandplay therapists in the geographic region in which the supervisee practices. The supervising sandplay therapist requires that prior to every consultation session, the supervisee prepare and transmit via e-mail detailed accounts of everything that transpired in the sandplay sessions to be discussed. These are transmitted without the use of encryption software. Electronic images of the trays are transmitted in an e-mail apart from the transmission of these reports. A savvy computer user would potentially be able to hack in, read these transmissions, and identify the client through identification of the supervisee's IP address and the identifying characteristics of the client that are contained in the transmission.

Case Example 2

A sandplay therapist takes photos of a client's sand trays with his or her iPhone or iPad and also maintains his or her calendar electronically on the same instrument. The iPhone/iPad records the date and time of the photo. The therapist uses only a four-digit password on the iPhone/iPad. A hacker would be able to access the material stored on the tablet or phone and associate the stored images with the client scheduled for a session on a particular date at a specified time. While the hacker may not understand the content of the sand tray image, he or she would be able to disseminate the images, together with the identity of the client, through other electronic media, such as Facebook or YouTube, potentially causing the client significant distress and embarrassment.

These heightened risks associated with cyber-supervision suggest that clients should be informed if such mechanisms are to be used for consultation–supervision and that they must provide informed consent to the transmission of their materials in this manner. We recommend that the following provisions be included in the informed consent form provided to the client if cyber-supervision is to be used, regardless of the nature of cyber-supervision, e.g., e-mail, chat room, televideoconferencing, etc:

1. The therapist will consult with a more experienced sandplay therapist during the course of the client's therapy.
2. All or part of the consultation-supervision will occur through electronic means, which may include one or more of the following: e-mail, chat room, instant messaging, televideoconferencing, Dropbox, Google Groups, or other means.
3. The therapist will use his or her best efforts to maintain the confidentiality of the information and the client's privacy.

4. Despite the therapist's best efforts, there remains a possibility that information may become known to others and, if this were to occur, neither the therapist nor the client would be able to control its further dissemination.
5. The client understands these risks and is willing to allow the use of cyber-supervision in conjunction with the therapist's provision of services to him or to her.

The therapist may also wish to include language that describes more fully the purpose of the supervision-consultation generally, the benefits of supervision for the client, that it is a routine practice within the mental health profession, and why the therapist must seek supervision that is not conducted face-to-face, e.g., geographic isolation, unavailability of a qualified sandplay supervisor within feasible travel distance. The client should also be made aware of how the therapist will address situations in which a breach may have occurred. The therapist will also want to apprise the client of possible alternatives to cyber-supervision, the risks associated with each such alternative, and whether the therapist is able and/or willing to provide services in the event that the client does not agree to cyber-supervision.

It is possible, of course, that the disclosure of such risks to the client will increase the likelihood that either the client will refuse consent or the client will choose to withhold information or provide inaccurate information to a greater extent than he or she might have otherwise done. Research has demonstrated that individuals often adjust the accuracy or completeness of disclosures to their mental health care providers to protect their privacy and confidentiality (California Healthcare Foundation 1999). However, a failure of the therapist to obtain client informed consent to cyber-supervision may not only represent a breach of trust and ethical violation but may also lead to legal consequences. The client might, for example, bring a civil lawsuit against both the supervisee and the supervisor for breach of confidentiality and/or violation of privacy even in the absence of hacking. The rights of a complainant or of a defendant in a civil case of cyber privacy infringement in the context of supervision appear to be unresolved at this moment (Shera 2014). This would seem to hold implications for professional indemnity insurance and possibly for professional licensure.

2.4.2   Supervisor–Supervisee Agreement

We also suggest that the supervisor and supervisee enter into a formal agreement if cyber-supervision is to be utilized. (This may be good practice even in situations involving only face-to-face supervision.) Like the client informed consent form, the agreement should specify that cyber-supervision is to be utilized and, in detail, explained what forms that will take, e.g., e-mail, televideoconferencing, and the general content that is expected to be transmitted, e.g., photos of sand trays, summaries of clinical notes. The risks associated with cyber-supervision should be noted as well. Additionally, the agreement should:

• Specify that the supervisor will use his or her best efforts to maintain client confidentiality and privacy and supervisee confidentiality and privacy;

- Describe the mechanisms that the supervisor has instituted to safeguard privacy and confidentiality, e.g., computer passwords, encryption, use of a VPN; and
- Describe the procedures that the supervisor will follow if he or she should become aware that a breach has occurred or may have occurred.

2.4.3   Supervisor Licensure and Scope of Practice Issues

The issue of licensure and scope of practice is both an ethical and a legal issue. Certainly, the provision of a consultation to a colleague outside of one's own country or across state lines in the USA would not be considered the provision of services outside of the scope of one's practice or practicing without a license. However, the circumstances are materially different if this consultation proceeds on an ongoing basis and is identified as a formal supervision relationship (Kanz 2001). In the USA, for example, it is unclear whether the supervisor would be subject to prosecution for practicing without a license in the state in which the supervisee is located, or subject to disciplinary action for practicing beyond the scope of his or her license in the state in which he or she practices, or both. Each sandplay therapist who provides cyber-supervision would want to investigate these issues with the authority(ies) who have granted their license(s).

2.5   Conclusions

Cyber-supervision offers significant benefits to both the sandplay therapist and the client. Indeed, it may be the only feasible mechanism by which to obtain supervision for sandplay therapists practicing in locales where there are no certified experienced sandplay therapists with similar client populations.

However, cyber-supervision involves substantial risks to confidentiality and privacy for both the client and the supervisee. The supervisor may also confront ethical and legal issues pertaining to the scope of practice and licensure. Initiation of a cyber-supervision relationship should not be made without knowledge of these risks and agreement to pursue cyber-supervision by all parties involved—the client, the supervisee, and the supervisor.

References

Barak, A. (1999). Psychological applications on the internet: A discipline on the threshold of a new millennium. *Applied & Preventive Psychology, 8*(4), 231–245.

Barnett, J. E. (2011). Utilizing technological innovations to enhance psychotherapy supervision, training, and outcomes. *Psychotherapy, 48*(2), 103–108.

Barnett, J. E., Cornish, J. E., Goodyear, R. K., & Lichtenberg, J.W. (2007). Commentaries on the ethical and effective practice of clinical supervision. *Professional Psychology: Research and Practice, 38,* 268–275.

California Healthcare Foundation. (1999). *Medical privacy and confidentiality survey: Summary and overview*. Oakland: California Healthcare Foundation. http://www.chcf.org/~/media/MEDIA%20LIBRARY%20Files/PDF/S/PDF%20survey.pdf. Accessed 11 May 2014.

Case Western Reserve University. (n.d.). Acceptable use policy: Frequently asked questions (FAQs). http://www.case.edu/its/policy/aup-faq.html. Accessed 11 May 2014.

Clingerman, T. L., & Bernard, J. M. (2004). An investigation of the use of e-mail as a supplemental modality for clinical supervision. *Counselor Education & Supervision, 44,* 82–95.

Falender, C. A., & Shafranske, E. P. (2004). *Clinical supervision: A competency-based approach*. Washington, D.C.: American Psychological Association.

Friedman, H. S., & Rogers Mitchell, R. (Eds). (2008). *Supervision of sandplay therapy*. New York: Routledge.

Glossbrenner, A. (1990). *The complete handbook of personal computer communications: The bible of the online world*. New York: St. Martin's Press.

International Society for Sandplay Therapy (ISST). (n.d.). Code of ethics. http://www.isst-society.com/homeng.php?site=ethics. Accessed 14 May 2014.

Internet World Stats. (2013). World internet usage and population statistics, June 30, 2012. http://www.internetworldstats.com/stats.htm. Accessed 11 May 2014.

Jerrell, J. M. (1983). Work satisfaction among rural mental health staff. *Community Mental Health Journal, 19,* 187–200.

Joinson, A. (1998). Causes and implications of disinhibited behavior on the internet. In J. Gackenbach (Ed.), *Psychology and the internet: Intrapersonal, interpersonal and transpersonal implications* (pp. 43–60). San Diego: Academic Press.

Kanz, J. E. (2001). Clinical-supervision.com: Issues in the provision of online supervision. *Professional Psychology: Research and Practice, 32*(4), 415–420.

Neukrug, E. S. (1991). Computer-assisted live supervision in counselor skills training. *Counselor Education and Supervision, 31,* 132–138.

Pew Research Center. (2014). Pew research internet project: Internet user demographics. http://www.pewinternet.org/data-trend/internet-use/latest-stats/. Accessed 11 May 2014.

Romans, J. S. C., Boswell, D. L., Carlozzi, A. F., & Ferguson, D. B. (1995). Training and supervision practices in clinical, counseling, and school psychology programs. *Professional Psychology: Research and Practice, 26,* 407–412.

Sandplay Therapists of America. (2012). Handbook of certified, teaching and practitioner member requirements and procedures for sandplay therapists of America (ISST). http://www.sandplay.org/pdf/STA_Handbook.pdf. Accessed 11 May 2014.

Schank, J. A. (1998). Ethical issues in rural counseling practice. *Canadian Journal of Counselling, 32,* 270–283.

Schultz, D. (2012). As patients' records go digital, theft and hacking problems grow. The Washington Post. http://www.kaiserhealthnews.org/ Stories/2012/June/04/electronic-health-records-theft-hacking.aspx. Accessed 11 May 2014.

Shera, R. (2014). Don't I know you?—Anonymity on the net. https://www.cs.auckland.ac.nz/~john/NetSafe/Shera.pdf. Accessed 13 May 2014.

Smith, R. C., Mead, D. E., & Kinsella, J. A. (1998). Direct supervision: Adding computer-assisted feedback and data capture to live supervision. *Journal of Marital and Family Therapy, 24,* 113–125.

Stebnicki, M. A., & Glover, N. M. (2001). E-supervision as a complementary approach to traditional face-to-face clinical supervision in rehabilitation counseling: Problems and solutions. *Rehabilitation Education, 15,* 283–293.

University of Florida Office of Academic Affairs. (1993). University of Florida faculty handbook. Gainesville: University of Florida Office of Academic Affairs. (Cited in D.E. Shapiro & C.E. Shulman. (1996). Ethical and legal issues in e-mail therapy. *Ethics & Behavior, 6*(2), 107–124, at 114.)

VandenBos, G. R., & Williams, S. (2000). The Internet versus the telephone: What is telehealth anyway? *Professional Psychology: Research and Practice, 31,* 490–492.

Wetchler, J. L., Trepper, T. S., McCollum, E. E., & Nelson, T. S. (1993). Videotape supervision via long-distance telephone. *American Journal of Family Therapy, 21,* 242–247.

Zuboff, S. (1988). *In the age of the smart machine: The future of work and power.* New York: Basic Books.

Chapter 3
Sandplay Therapy and Access to Mental Health Care Services: Present Barriers and Future Promise

Sana Loue

3.1 Introduction

Research has consistently found that members of minority racial/ethnic groups are significantly less likely than their nonminority counterparts to access or utilize mental health care. This is as true both in the USA and in other regions of the world (Bebbington et al. 2000; Carta et al. 2005; ESEMeD/MHEDEA 2000 2004; Lindert et al. 2008; Pomare 1980; Pomare and de Boer 1988; Pomare et al. 1995; Szczepura 2005; Sundquist 2001). Research conducted in the USA has found that when minority individuals are able to access care, it is likely to be of lesser quality (Garland et al. 2005; Harris et al. 2005; Institute of Medicine 2003; United States Department of Health and Human Services 2001; Wells et al. 2001), leading to a disproportionate burden of disability attributable to mental illness (United States Department of Health and Human Services 2003). These disparities in access are increasingly of concern to health care professionals and policy makers (Institute of Medicine 2003).

The ethical principle of distributive justice has been explained as "the fair and equitable allocation of burdens and privileges, rights and responsibilities, and pains and gains in society" (Prilleltensky 2012, p. 1). Indeed, various mental health professions specifically recognize the principle of justice in their ethical codes and guidelines. For example, the *Ethical Principles of Psychologists and Code of Conduct* of the American Psychological Association provides:

> Psychologists recognize that fairness and justice entitle all persons to access to and benefit from the contributions of psychology and to equal quality in the processes, procedures, and services being conducted by psychologists. Psychologists exercise reasonable judgment and take precautions to ensure that their potential biases, the boundaries of their competence, and the limitations of their expertise do not lead to or condone unjust practices.
> (American Psychological Association 2010, 3–4, Principle D)

S. Loue (✉)
School of Medicine, Case Western Reserve University, Cleveland, OH, USA
e-mail: sana.loue@case.edu

© Springer International Publishing Switzerland 2015

S. Loue (ed.), *Ethical Issues in Sandplay Therapy Practice and Research,*
SpringerBriefs in Social Work, DOI 10.1007/978-3-319-14118-3_3

Similarly, the *APS Code of Ethics* of the Australian Psychological Society states:

> They [psychologists] have a high regard for the diversity and uniqueness of people and
> their right to linguistically and culturally appropriate services. *Psychologists* acknowledge
> people's right to be treated fairly without discrimination or favouritism, and they endeavour
> to ensure that all people have reasonable and fair access to *psychological services* and share
> in the benefits that the practice of psychology can offer. (Australian Psychological Society
> Limited 2007, p. 11) (emphasis in original)

Codes of ethics and ethical guidelines for social workers contain similar provisions
relating to the principle of justice. The Clinical Social Work Association states in
its *Code of Ethics:*

> Clinical social workers recognize a responsibility to participate in activities leading toward
> improved social conditions. They should advocate and work for conditions and resources
> that give all persons equal access to the services and opportunities required to meet basic
> needs and to develop to the fullest potential. (Clinical Social Work Association 2006)

This chapter uses the USA as a case example to illustrate the structural barriers confronting individuals who are poor and/or are members of minority groups in their
efforts to access mental health care services. The chapter then proceeds to explore
the structural barriers that face both such clients who may be seeking sandplay therapy specifically and the mental health care providers who serve them. The chapter
concludes with recommendations for structural modifications related to sandplay
certification and membership that may increase access to sandplay by minority and
poverty-level clients and facilitate the ability of their mental health providers to
participate in the sandplay therapy profession.

This examination of the profession of sandplay therapy is in no way meant to
imply that sandplay therapy as a profession or sandplay therapists as individual
practitioners are or should be expected to solve the societal problem of access to
care to mental health care for minority or poor individuals. Indeed, it would be the
height of arrogance to suggest that sandplay therapists, either individually or as an
organization, would be able to do so. However, because structural factors may pose
a barrier to accessing mental health care (Cabassa et al. 2006), the ethical principles
that govern our therapeutic professions suggest that an examination of the structural
elements of sandplay therapeutic practice is called for. In so doing, it is important
to recognize that such structural barriers are not unique to sandplay therapy as a
therapeutic modality. Indeed, they appear to characterize various specialized modalities, such as Freudian and Jungian psychoanalysis. Further, this is not to suggest
that even if all structural barriers to accessing sandplay were to be eliminated, that
sandplay therapy would be an appropriate modality for all use with all clients.

3.2 Structural Barriers in Accessing Mental Health Services: The USA as a Case Study

In the USA, a lack of health care insurance has been implicated as a major factor in
the ability of individuals to obtain needed mental health care (Cunningham 2009).
This is an even greater difficulty for minority individuals (Albizu-Garcia et al.

2001; Cunningham 2009; McAlpine and Mechanic 2000; Snowden 2001; Vega and López 2001; Wells et al. 1989). Even among individuals with health care insurance, African Americans of any economic status may have fewer financial resources than their non-Hispanic White counterparts. As a result, the cost sharing associated with their health care insurance may present a greater burden to them than to non-Hispanic Whites in the same income bracket (Alegría et al. 2002).

Individuals' ability to speak English also plays a major role in their ability to access mental health services, independent of the effect of race/ethnicity. Several studies have found that Latino and Asian Pacific Islander individuals who do not speak English or are bilingual are less likely to receive mental health services than those who speak only English (Loue 2005, 2011; Sentell et al. 2007), even among those who have health care insurance (Sentell et al. 2007). Mental health services for Latinos would be greatly improved if there were a greater number of bilingual bicultural mental health care providers (United States Department of Health and Human Services 2001).

Poverty in particular has been implicated as one of the primary reasons for minority individuals' inability to access mental health care services. African Americans and Latinos are more likely to live in poverty than non-Hispanic Whites. The circumstances in which they often live due to poverty also negatively affect their access. Individuals with mental illness and members of minority racial/ethnic groups are disproportionately concentrated in neighborhoods characterized by high levels of poverty (Dear and Wolch 1987; Faris and Dunham 1960; Srole and Fischer 1962; Wolch and Dear 1994), unemployment, residential turnover, crime, homelessness, and substance use (Sampson et al. 1997; Wilson 1987), and low levels of service providers apart from public hospitals and mental health centers (Lewin and Altma 2000; Snowden 1999). An analysis of data drawn from the 1990–1992 National Comorbidity Survey that included information about 8098 English-speaking adults aged 15–54 years found that Latino families with an income of less than US$ 15,000 per year were less likely than equally poor non-Latino Whites to have access to mental health care services, even after considering other factors such as insurance status and psychiatric comorbidities (Alegría et al. 2002). The researchers also found that both African Americans and Latinos were less likely to use mental health services than non-Latino Whites, even when the analysis considered income and geographic region of residence. These findings lend support to the idea that the circumstances of poverty, and not poverty alone, may contribute to difficulties accessing and utilizing mental health care.

Children living in such neighborhoods may be especially vulnerable to mental health problems as a result of their exposure to high levels of violence, unsafe housing, and sense of isolation (Aneshensel and Sucoff 1996; Garbarino 1992). Children at greatest risk for psychopathology and those with the most difficult environmental situations may be the least likely to be engaged in mental health care (Cohen and Hesselbart 1993; Kazdin 1993; McKay et al. 1996). These circumstances have prompted some researchers to recommend that "mental health services must be tailored to meet the unique needs of minority racial/ethnic groups in different community settings" (Chow et al. 2003). Others have argued somewhat similarly that mental health practitioners providing care to minority urban children "must incorporate in their treatment...an understanding of how multiple socio-environmental factors...interact with individual and family dynamics within a specific ethnic cultural-racial context" (González 2005).

It is also important for individuals to know where they can obtain the needed mental health services in order to access them. For example, Mexican Americans who know where they can obtain mental health services are more likely to utilize them compared to their counterparts who do not know where to find such care (Vega and López 2001). This has also been found to be the case among Puerto Ricans (Ortega and Alegría 2002).

These research findings have practical implications for the initiation, establishment, and continuation of a therapeutic relationship with a mental health care provider, regardless of whether that provider offers sandplay therapy. First, if an individual does not know where services can be obtained, he or she will not be able to access the care even if a need is recognized and the service is desired. Language may prove to be a barrier to locating a venue that offers the needed service and/or to identifying a specific provider who is able to communicate in a language that the individual can understand and use. Then, even if the individual has this information, care may be out of reach financially due to a lack of insurance and/or copays and deductibles may render it prohibitively expensive.

Assume that the individual has overcome these initial barriers and has initiated therapy. The individual's attendance at prescheduled therapy sessions may be inconsistent and sporadic for any number of reasons: the unavailability of funds for public transportation, gasoline, or car repairs, so that the individual cannot get there; a demand from the person's employer that he or she work at that time; a child's illness and the unavailability or unaffordability of child care; external circumstances, e.g., neighborhood violence, that effectively prevents the individual from safely leaving his or her house. Not infrequently, care providers may become frustrated and the client may come to be seen as unreliable and nonadherent to treatment. And yet, if mental health care providers are to provide services to individuals living in such circumstances, they must tailor their services appropriately "to meet the unique needs of [individuals living in such] community settings" (cf. Chow et al. 2003).

Case Example

A young African American transgender woman (male-to-female) "pops in" to a local drop-in center to see the on-call therapist who is there during evening hours. She explains that she has never seen a therapist prior to that time. She does not offer a clear explanation as to why she is seeking services at that time. She confides that she self-identifies as a lesbian and that she does not divulge that information to many people. She more often identifies herself to others as a "straight." She also confides that she has been cutting and shows the therapist the scars that run up her right arm, now covered with a long-sleeved t-shirt. At first skeptical about sandplay, she quickly becomes engaged in the process once she begins. As she places figures in the sand, she explains how each represents a member of her family. She tells the therapist that her living situation has deteriorated since her mother's new boyfriend joined the household and she is often the focus of his verbal attacks. She has recently moved in with friends, who have been pressuring her to engage in sex work in order to pay her share of the rent. She sobs as she tells this to the therapist but, at the conclusion of the session, says that she feels much better because it is the first time that she was able to share these experiences and her feelings openly. She makes an appointment to see the therapist on the same day at the same time the following week. She does not, however, appear at the drop-in center for several weeks. On her return several weeks later and without an appointment, she advises the therapist that she was in jail. She had been picked up by the police for solicitation and had been unable to post bail.

That said, it is not at all clear that providers offering sandplay therapy to clients in such circumstances are able to do so and receive recognition within their respective professional sandplay societies as sandplay therapists. The following portion of this chapter explores the structural barriers within the sandplay therapy profession to achieving such recognition and the potential implications for their clients.

3.3 Sandplay Therapy: Current Barriers

3.3.1 International and National Membership Requirements

The International Society of Sandplay Therapists (ISST) is an international umbrella society that operates through national sandplay organizations located in numerous countries throughout the world. ISST has established three categories of membership: (1) national societies, (2) individuals who are not members of a national society and who practice in a geographical area where there are fewer than the five minimum ISST-certified sandplay therapists to establish a national society or who have been granted an exception by the ISST Board, and (3) honorary members who, in general, are the founding members of the society. In general, ISST recognizes only one national society per country. Certified members of a national society are indirect members of ISST (International Society of Sandplay Therapy 2006).

The Sandplay Therapists of America (STA) is the ISST-recognized national society in the USA. STA recognizes three classes of members: certified sandplay therapists, certified sandplay therapist-teaching members, and sandplay practitioners. Individuals who do not meet the standards delineated for any of these three classes may join the organization as associates. Associates need not be licensed mental health care professionals and those who are therapists may or may not offer sandplay therapy within their practice.

Like many professional societies, individuals seeking recognition as professional members, in this case certified members, teaching members, or sandplay practitioners, must satisfy various preliminary requirements prior to the application for membership. Specifically, STA requires a graduate degree in a helping profession such as medicine, psychology, certified social work, pastoral counseling, school counseling, or marriage and family counseling from a regionally accredited university; demonstration of a clinical knowledge base that includes knowledge of psychotherapy, psychodiagnosis, and psychology obtained through formal study and 2 years of supervised certified experience; a license or credential to practice in profession; completion of 2000 h of direct supervised certified or counseling experience; and evidence of in-depth inner development and insight gained through analysis and/or psychotherapy that must have occurred within the 10 years immediately prior to the date of application (Sandplay Therapists of America 2013).

In addition to these prerequisites, STA's *Handbook of Practitioner Member Requirements and Procedures* (2013) requires that therapists seeking certification as a full member undergo a personal sandplay process consisting of a minimum of 40 sessions; complete a comprehensive program of study of 120 h or more that provides an introduction to sandplay therapy, Jungian theory, and symbolism and clinical sandplay practice; receive a minimum of 80 h of consultation (individual and group) with an ISST-certified member; and author two preliminary papers and a final case study that are reviewed and approved by the requisite number of fully certified sandplay therapists (Sandplay Therapists of America 2013). Analysis and/ or psychotherapy may be concurrent with personal sandplay process. The combined total number of hours of analysis, psychotherapy, and personal sandplay process is a minimum of 100 h. These certification requirements are somewhat modified for individuals who have already achieved status as a Jungian therapist at the time of their application for certified membership. Teaching membership requires fulfillment of criteria in addition to those noted above.

Membership requirements are ostensibly more relaxed for mental health care professionals who are seeking recognition by the association of their expertise as sandplay therapists, but are not seeking full certification. Recognition as a sandplay practitioner requires:

- A valid state license or credential as a mental health professional or a professional license, credential, certificate, or equivalent in an allied field, such as nursing, teaching, or spiritual direction;
- A commitment to in-depth inner development and insight as gained through analysis and/or psychotherapy;
- Completion of a sandplay process with an STA/ISST Certified Sandplay Therapist (CST) or Certified Sandplay Therapist-Teacher (CST-T);
- Completion of a minimum of 36 h of education in sandplay with a CST-T or at an STA-sponsored conference, seminar, or workshop, including 18 h of an introductory course in sandplay, 12 h of which may be earned through field-tested, STA-approved on-line courses;

- Participation in group consultation with a CST-T for a minimum of 25 sessions in which the applicant presents at least 5 h of sandplay case material or in individual sandplay consultation for a minimum of 15 h or in a combination of group and individual consultation sessions for a total of 20 sessions. If an individual applicant selects a combination of individual and group consultation sessions, at least 2 h of every 10 h of group consultation must be presentation hours by the applicant.
- Consultation from someone other than the therapist with whom they completed their sandplay process; and
- Work with a minimum of three clients or students per week, who engage with sandplay on a regular basis, for a minimum of 1 year (9 months for school counselors) under the consultation of a CST-T.

A comparison of the requirements for recognition as a sandplay practitioner with those required for full certification suggests that these more relaxed requirements for sandplay practitioner may actually be equally or even more demanding than those required for full certification. For example, status as a sandplay practitioner requires that the therapist provide sandplay therapy on an ongoing basis to a minimum of three clients per week for a minimum period of 9–12 months. In contrast, the requirements for certification make no mention of a minimum number of clients or of the frequency or duration of the therapy provided. The educational requirement for each status is similar, as is the requirement that the applicant complete a personal sandplay process. Potentially, this may entail fewer hours for the sandplay practitioner-applicant than the certification-applicant, if only because the *Handbook* does not specify the number of sessions required for the sandplay practitioner-applicant.

3.3.2 Implications for Access to Services and Associated Ethical Issues

Regardless of which route an individual chooses to take to gain expertise in sandplay therapy—certification or sandplay practitioner—the commitment requires a significant investment of time and money. Both the hours required for the personal sandplay process and the consultation hours will be recognized only if they are done with a certified STA teaching member. Although some members are willing to provide these services on a sliding scale, many are not; depending on the geographical region, the average rate per hour is in the range of US$ 100–150. Forty hours of personal process and 80 h of consultation at US$ 100 per hour results in an investment of US$ 12,000. Some STA teaching members are willing and able to provide consultation via Skype, but others are not. (For a discussion of the ethical issues associated with the provision of consultation via Skype, see Chap. 2 in this volume.) In situations in which individuals must travel to obtain their consultation and/or personal process hours, travel costs represent an additional expense. Clearly, cost may be a significant barrier for individuals who are just beginning their private practice, who work with social service agencies that have limited budgets for staff professional development and travel, and those who provide therapy on a sliding or gratis basis to clients with limited financial means.

The relationship between the requirements for recognition as a sandplay practitioner or certified member and the context in which therapy is provided must also be considered. Both the *Handbook of Certified, Teaching and Practitioner Member Requirements and Procedures for Sandplay Therapists of America (ISST)* and discussions at the annual meetings of the STA strongly suggest that individuals cannot and should not receive either certification or recognition as a sandplay practitioner unless they have provided sandplay therapy to clients who have themselves completed their sandplay process. This requirement ignores the context in which some clients live and the context in which some therapists practice.

One must ask, then, why a therapist, whose practice is new, focuses on financially marginalized individuals or is situated in a social service agency which would choose to invest the time and money in the requisite training if he or she may never achieve full professional recognition within the sandplay association. The structural barriers erected by the requirements for certification/sandplay practitioner status have the potential effect of reducing diversity within the organization and the sandplay profession and of eliminating or greatly reducing the availability of sandplay therapy to clients who would potentially benefit from this modality—a result that this author argues contravenes the ethical principle of justice.

3.4   Sandplay Therapy: Future Promise

Remediation of the structural barriers that potentially limit both diversity within the sandplay profession and client access to sandplay therapy will require a multidimensional approach that targets modifiable structural elements (cf. Cabassa et al. 2006). Although STA has instituted a few initiatives to increase diversity within the profession, these efforts have been somewhat limited and their impact on client access has not been assessed.

Scholarships are currently available for attendance at the STA annual meetings to individuals working with diverse or limited-income populations or who are themselves members of a minority group or of limited means. To date, scholarships have been provided in the absence of supporting documentation apart from an applicant's own statement. Additionally, recipients are not obligated to provide feedback to the organization regarding their use of sandplay with clients or the continuation of their work with the client population. Consequently, it is unclear to what extent these scholarships actually enhance the diversity within the profession or client access to the modality.

Exceptions to the requirements for certification/sandplay practitioner status are ostensibly available to individuals through an application to the Exceptions Committee. The parameters of these exceptions are not public, so that it is unclear whether an exception in these circumstances has ever been or will ever be granted. The lack of transparency regarding this process may itself be a barrier to application.

Additionally, a true commitment to expansion of client access and membership diversity suggests that a revision of the requirements is necessary, rather than the trickle-down effect of individual applications for exceptions. Revision of the requirements for certification/sandplay practitioner status clearly requires careful consideration in order to ensure competency of the individuals certified or recognized as sandplay practitioners.

Various professional organizations have sought to motivate their members to provide pro bono/gratis services to individuals and nonprofit organizations that are unable to otherwise afford them. Notable examples include the American Bar Association and state bar associations, which not only try to motivate their members to provide such services but also have established accounts for the receipt of member donations to support such services. The development of a similar aspirational goal by the STA/ISST for its members could lead to increased client diversity and expanded client access to sandplay therapy.

Increased diversity of both sandplay therapists and the client population accessing sandplay therapy will require additional, broader efforts to disseminate information about this treatment modality. The STA may wish to consider using social marketing techniques to inform prospective members and clients in addition to its current efforts. If such efforts are to be made, it is critical that ISST/STA align their requirements for professional recognition with the practice realities faced by the therapists and clients to whom these marketing efforts are addressed.

3.5 Conclusions

Minority and uninsured/underinsured individuals often lack access to adequate mental health services, including sandplay therapy. The requirements established by the national and international professional certifying organizations for sandplay inadvertently erect structural barriers that ignore the lived realities of many potential clients and the contexts in which their mental health care providers practice. These barriers may ultimately lead to a decrease in the diversity of both sandplay therapists and clients. Such results appear to contravene the ethical principle of distributive justice, i.e., "the fair and equitable allocation of burdens and privileges, rights and responsibilities, and pains and gains in society" (Prilleltensky 2012, p. 1). Consideration of structural reforms to membership requirements is critical if this is to be remedied.

References

Albizu-Garcia, C. E., Alegría, M., Freeman, D., & Vera, M. (2001). Gender and health services use for mental health problems. *Social Science & Medicine, 53,* 865–878.

Alegría, M., Canino, G., Ríos, R., Vera, M., Calderón, J., Rusch, D., & Ortega, A. N. (2002). Inequalities in use of specialty mental health services among Latinos, African Americans, and non-Latino Whites. *Psychiatric Services, 53*(2), 1547–1555.

American Psychological Association. (2010). Ethical principles of psychologists and code of conduct. http://www.apa.org/ethics/code/principles.pdf. Accessed 4 Aug 2014.

Aneshensel, C.S., & Sucoff, C.A. (1996). The neighborhood context of adolescent mental health. *Journal of Health and Social Behavior, 37,* 293–310.

Australian Psychological Society Limited. (2007). *APS code of ethics.* http://www.psychology.org. au/Assets/Files/APS-Code-of-Ethics.pdf. Accessed 4 Feb 2015.

Bebbington, P. E., Meltzer, H., Brugha, T. S., Farrell, M., Jenkins, R., Ceresa, C., et al. (2000). Unequal access and unmet need: Neurotic disorders and the use of primary care services. *Psychological Medicine, 30,* 1359–1367.

Cabassa, L. J., Zayas, L. H., & Hansen, M. C. (2006). Latino adults' access to mental health care: A review of epidemiological studies. *Administration and Policy in Mental Health and Mental Health Services Research, 33*(3), 316–330.

Carta, M. G., Bernal, M., Hardoy, M. C., Haro-Abad, J. M., & The Report on the Mental Health in Europe Working Group. (2005). Migration and mental health in Europe ("the state of the mental health in Europe" working group: Appendix I). *Clinical Practice and Epidemiology in Mental Health, 1,* 13.

Chow, J. C.-C., Jaffee, K., & Snowden, L. (2003). Racial/ethnic disparities in the use of mental health services in poverty areas. *American Journal of Public Health, 93*(5), 792–797.

Clinical Social Work Association. (2006). Code of ethics. http://www.clinicalsocialworkassociation.org/about-us/ethics-code. Accessed 4 Aug 2014.

Cohen, P., & Hesselbart, C. S. (1993). Demographic factors in the use of children's mental health services. *American Journal of Public Health, 83,* 49–52.

Cunningham, P. J. (2009). Beyond parity: Primary care physicians' perspectives on access to mental health care. *Health Affairs, 28*(3), w490–501. Epub 2009 Apr 14.

Dear, M., & Wolch, J. (1987). *Landscapes of despair.* Princeton: Princeton University Press.

ESEMeD/MHEDEA 2000. (2004). Use of mental health services in Europe: Results from the European study of the Epidemiology of Mental Disorders (ESEMeD) project. *Acta Psychiatrica Scandinavica, 109*(Suppl. 420), 47–54.

Faris, R., & Dunham, H. W. (1960). *Mental disorders in urban areas: An ecological study of schizophrenia and other psychoses.* New York: Hafner Publishing Co.

Garbarino, J. (1992). *Children and families in the social environment* (2nd ed.). New York: Aldine De Gruyter.

Garland, A. F., Lau, A. S., Yeh, M., McCabe, K. M., Hough, R. L., & Landsverk, J. A. (2005). Racial and ethnic differences in utilization of mental health services among high-risk youths. *American Journal of Psychiatry, 162*(7), 1336–1343.

González, M. J. (2005). Access to mental health services: The struggle of poverty affected urban children of color. *Child and Adolescent Social Work Journal, 22*(3–4), 245–256.

Harris, K. M., Edlund, M. J., & Larson, S. (2005). Racial and ethnic differences in the mental health problems and use of mental health care. *Medical Care, 43*(8), 775–784.

Institute of Medicine. (2003). *Unequal treatment: Confronting racial and ethnic disparities in healthcare.* Washington D.C.: National Academies Press.

International Society of Sandplay Therapy. (2006). Statutes. http://www.isst-society.com/homeng. php?site=statutes. Accessed 11 Feb 2014.

Kazdin, A. E. (1993). Premature termination for treatment among children referred for antisocial behavior. *Journal of Child Psychology and Psychiatry and Allied Disciplines, 31,* 415–425.

Lewin, M., & Altma, S. (2000). *America's health care safety net: Intact but endangered.* Washington D.C.: National Academy Press.

Lindert J., Schouler-Ocak M., Heinz A., & Priebe, S. (2008). Mental health, health care utilization of migrants in Europe. *European Psychiatry, 23,* 814–820.

Loue, S. (2011). *"My Nerves Are Bad" ("Mis Nervios Estan Malos"): Puerto Rican women managing mental illness and HIV risk.* Nashville: Vanderbilt University Press.

Loue, S., & Mendez, N. (2005). Barriers to treatment for severely mentally ill Latinas. Presented at the All-Ohio Institute on Community Psychiatry. Cleveland, 18–19 March 2005.

McAlpine, D. D., & Mechanic, D. (2000). Utilization of specialty mental health care among persons with severe mental illness: The roles of demographics, need, insurance, and risk. *Health Services Research, 35*(1), 278–292.

McKay, M., McCadam, K., & Gonzalez, J. (1996). Addressing the barriers to mental health services for inner-city children and their caregivers. *Community Mental Health Journal, 32,* 353–361.

Ortega, A. N., & Alegría, M. (2002). Self-reliance, mental health need, and the use of mental healthcare among island Puerto Ricans. *Mental Health Services Research, 4*(3), 131–140.

Pomare, E. W. (1980). Hauora: Maori standards of health: A study of the 20-year period 1955–1975. Special Report Series No. 7. Auckland: Medical Research Council of New Zealand.

Pomare, E. W., & de Boer, G. M. (1988). Hauora: Maori Standards health; a study of the years 1970–1984. Special report series 78. Wellington: Department of Health and Medical Research Council.

Pomare, E., Keefe-Ormsby, V., Ormsby, C., Pearce, N., Reid, P., Robson, B., & Watene-Haydon, N. (1995). *Hauora Māori standards of health III*. Wellington: Te Ropu Rangahau Hauora a Eru Pomare Wellington School of Medicine.

Prilleltensky, I. (2012). Wellness as fairness. *American Journal of Community Psychology, 49*(1/2), 1–21.

Sampson, R., Raudnbush, S., & Earls, F. (1997). Neighborhoods and violent crime: A multilevel study of collective efficacy. *Science, 277,* 918–924.

Sandplay Therapists of America. (2013). Handbook of certified, teaching and practitioner member requirements and procedures for Sandplay Therapists of America (ISST). http://www.sandplay.org/pdf/STA_Handbook.pdf. Accessed 14 Feb 2013.

Sentell, T., Shumway, M., & Snowden, L. (2007). Access to mental health treatment by English language proficiency and race/ethnicity. *Journal of General Internal Medicine, 22*(Suppl. 2), 289–293.

Snowden, L. (1999). Psychiatric inpatient care and ethnic minority populations. In J. Herrera, W. Lawson, & J. Sramek (Eds.), *Cross cultural psychiatry* (pp. 261–274). New York: Wiley.

Snowden, L. (2001). Barriers to effective mental health services for African Americans. *Mental Health Services Research, 3*(4), 181–187.

Srole, L., & Fischer, A. (1962). *Mental health in the metropolis: The Midtown Manhattan study*. New York: McGraw-Hill.

Sundquist, J. (2001). Migration, equality and access to health care services. *Journal of Epidemiology & Community Health, 55,* 691–692.

Szczepura, A. (2005). Access to health care for ethnic minority populations. *Postgraduate Medicine Journal, 81,* 141–147.

United States Department of Health and Human Services. (2001). *Mental health: Culture, race, and ethnicity: A supplement to mental health: A report of the Surgeon General.* Rockville: United States Department of Health and Human Services.

United States Department of Health and Human Services. (2003). *The President's New Freedom Commission on Mental Health: Achieving the promise: Transforming mental health care in America.* Rockville: United States Department of Health and Human Services. [DHHS Pub. No. SMA-03-3831].

Vega, W. A., & López, S. R. (2001). Priority issues in Latino mental health services research. *Mental Health Services Research, 3*(4), 189–200.

Wells, K. B., Golding, J. M., Hough, R. L., Burman, M. A., & Karno, M. (1989). Acculturation and the probability of use of health services by Mexican Americans. *Health Services Research, 24*(2), 237–257.

Wells, K., Klap, R., Koeke, A., & Sherbourne, C. (2001). Ethnic disparities in unmet need for alcoholism, drug abuse, and mental health care. *American Journal of Psychiatry, 158*(12), 2027–2032.

Wilson, W. (1987). *The truly disadvantaged: The inner city, the underclass, and public policy.* Chicago: University of Chicago Press.

Wolch, J., & Dear, M. (1994). *Malign neglect: Homelessness in an American city.* San Francisco: Jossey-Bass.

Chapter 4
Transference and Countertransference in an Ethical Context

Sana Loue

4.1 Introduction

Issues of transference and countertransference are generally conceived of as clues to the development of a better understanding of the client and a deeper self-awareness on the part of the therapist. The *Social Work Dictionary* (Barker 2003, p. 439) defines transference as "emotional reactions that are assigned to current relationships but originated in earlier, often unresolved and unconscious experiences."

Conversely, countertransference refers to the "conscious or unconscious emotional reactions to a client experienced by [the therapist]..." (Barker 2003, p. 100). Transference and countertransference, which can be positive or negative, can be used by the therapist with the client to further the client's resolution of past conflicts. Countertransference may also provide the therapist with additional insights into the client and the client's effect on others around him or her (Racker 1968). Similarly, the therapist can explore his or her reactions through consultation/supervision with another therapist. This is an important component of one's practice.

In sandplay, we understand the concept of *co-transference*, which comprises both transference and countertransference. Bradway (1991, p. 29) explained that these feelings are

necessarily determined by earlier and current happenings. They are, of course, both positive and negative, conscious and unconscious. And it is not just the person coming for therapy who projects; the therapist does also. Both may find hooks in the other in which to project, or hang, the unused parts of themselves, or repressed parts, or personal images from the past, or archetypal images. And both respond to these projections. One can't help but be affected by the projections of significant others. Moreover, both projections and responses are often entirely at an unconscious level. The therapeutic relationship is a mix; a complex mix; a valuable mix. It is to this mix that I am referring when I use the term co-transference.

S. Loue (✉)
School of Medicine, Case Western Reserve University, Cleveland, OH, USA
e-mail: sana.loue@case.edu

© Springer International Publishing Switzerland 2015
S. Loue (ed.), *Ethical Issues in Sandplay Therapy Practice and Research,*
SpringerBriefs in Social Work, DOI 10.1007/978-3-319-14118-3_4

Bradway and McCoard (1997, p. 34) explained how the concept of co-transference differs from transference and countertransference, in that the concept of co-transference elicits

> a feeling with (*co*), rather than a feeling against (*counter*). I use the term co-transference to designate the therapeutic feeling relationship between therapist and patient. These inter-feelings seem to take place almost simultaneously, rather than sequentially as the composite terms transference-countertransference suggests.

However, issues of co-transference, when not addressed appropriately, may also give rise to ethical issues related to practice competence. A failure to recognize and/or address issues of transference and/or countertransference appropriately could potentially subject a client to a risk of harm. The *Ethical Principles of Psychologists and Code of Conduct* advises psychologists to "take reasonable steps to avoid harming their clients/patients ... and to minimize harm where it is foreseeable and unavoidable" (American Psychological Association 2010, § 3.04). The National Association of Social Workers (NASW) Code of Ethics provides that "[s]ocial workers' primary responsibility is to promote the wellbeing of clients" (National Association of Social Workers 2008, § 1.01). One must question whether this is possible if the social worker–therapist does not recognize and/or address issues of co-transference.

Supervision provides therapists with an opportunity to gain further insight into themselves, their client, and the dynamics between them. It ultimately may help to increase the therapist's effectiveness with the client and reduce the possibility of a harmful therapist–client dynamic. From a Jungian perspective, supervising sandplay is the supervision of the unconscious and the creative imagination (Friedman and Mitchell (2008, p. 2). Supervision in the context of sandplay

> can also be very rewarding because it enlarges the psychic and spiritual horizon of all participants and develops a beautiful intercultural understanding and awareness between them about their own and others' sociocultural origins. The human psyche, or the *mysterium,* as we Jungians dare to call the soul, is so much wider and deeper and older than we could ever imagine (Amman 2008, p. 108).

This chapter focuses on issues of co-transference that may arise in working with sandplay clients whose trays evidence spiritual elements and potential associated ethical issues. I focus in this chapter on co-transference as it relates to spiritual issues because I am ordained as an interfaith minister in addition to being a licensed independent social worker (LISW). I tend to see spiritual issues and elements even in everyday occurrences and manifestations. When I work with clients, I often feel that there is a third presence that joins me and the client, guiding the client in his or her search and assisting me to gain an understanding of the client.

I begin this chapter with a brief presentation of two clients and their sand trays. I follow this with a "View from the Therapist," an explanation of how I believe the sandplay process comprises sacred dimensions. I follow this with a discussion of how inadequate attention to what I bring to the therapy process with these two clients could disrupt their processes if I were unaware and did not seek supervision.

4.2 The Clients

4.2.1 Dora

An older woman, who I will call Dora, consulted with me on only one occasion for her symptoms of depression. Dora was in her late middle-age years at the time I met her. Her children were all grown, with their own families, and she did not have a partner. She had recently been released from inpatient care for her depression and, because of the severity of her depression, was enrolled in an intensive outpatient program and was following a regimen of antidepressants prescribed by her psychiatrist. Dora had suffered from severe depression during varying periods of her life. Both her psychiatrist and psychologist believed that Dora had no knowledge of her son's cancer diagnosis and poor prognosis for recovery; no one had conveyed this information to her for fear that her depression would worsen.

Dora had never seen a sand tray before she came into my office. At first reluctant and somewhat embarrassed about her interest in the tray, she decided, without explanation, to give it a try.

In making her tray, Dora placed figures of the Madonna and an angel in the center of a circle formed by four girls. Dora said nothing about the Madonna. Although not referring to the Madonna at all, Dora talked extensively about how the angels are there to protect the young children: "The angels are there to look after the children. They are innocent."

I obtained Dora's permission to share her sand tray with her psychiatrist and psychologist. I believed that on some level, Dora had surmised that her son was ill and possibly dying. Perhaps the tray was her way of creating an opening to discuss her fears and worries. I suggested this to the psychiatrist and psychologist as I shared her tray with them. They disagreed and continued to insist that Dora was not to be told about her son's illness and the likelihood that he would soon die.

4.2.2 Dustin

We can consider the second sand tray made by a client who I will call Dustin. As he made this tray, seen below, it seemed as though he was acting out a movie of his life, choosing figures that he said represented each of the characters in his own life.

Dustin indicated that the figure toward the right center, hands on hips, represented his grandmother, who he said frequently harangued him. He placed the small soldier, rifle in hand, directly in front of the grandmother's figure, indicating that this represented himself. The outstretched figure in the upper right corner, Dustin said, was his mother: "Nothing is going to get her to stand up." He surrounded this figure with small Viking figures that Dustin said represented some of his other children. The ape bedside her with its hand over her crotch, he said, was her most recent boyfriend. Dustin said the small teddy bear near the two figures represented his youngest brother. Dustin placed several weapon-carrying soldier figures and a wrestler in the upper left-hand corner of the tray, remarking that those figures represented his uncle and cousin who had sexually abused him, together with some other cousins. Dustin placed the pink-haired troll in the front center of the tray in such a way that it was looking at the unfolding scene in the tray. Dustin said that this figure represented his aunt, who was "always wanting to be in control, as if she is directing a play" and "like the conductor in an orchestra who says who will play when."

Dustin selected a tall, muscular wrestling-type figure with clenched fists and placed it on its back outside of the sand tray on the floor. He said this represented his grandfather who was a "deadbeat," "laying like that drunk." Dustin then selected two house-like figures. He placed the smaller of the two on the ledge of the tray, indicating that this small house represented the house where the sexual abuse had occurred. The larger "house," which was actually a schoolhouse, was placed toward the lower left center of the tray. He placed a Buddha figure next to the schoolhouse figure; he appeared to be unaware that this house was a schoolhouse or that the figure that he had placed nearby was a Buddha.

I have written previously about Dustin and his short-lived sandplay experience (Loue 2009). The figure of the Buddha often symbolizes a teacher or healer on a journey toward inner wisdom; the term "Buddha" itself means "awakened" (Smith 1991). Is it possible that Dustin's selection of the Buddha figure and his placement of it by a schoolhouse reflected the commencement of his own journey toward self-knowledge, integration, and healing? Might he have been looking for his own Buddha nature?

Dustin, sighing as he completed this sandplay, said, "It's like outside looking in when I look back, like sleeping with my eyes open."

4.3 "View from the Therapist": Seeing the Sacred in Sandplay

What does it mean to be sacred? How is sandplay sacred?

To be sacred is to be
1. Dedicated to or set apart for the worship of a deity. 2. Worthy of religious veneration …. 3. Made or declared holy …. 4. Dedicated to or devoted exclusively to a single use, purpose, or person …. 5. Worthy of respect; venerable. 6. Of or relating to religious objects, rites, or practices … (*The American Heritage Dictionary of the English Language* 2000, p. 1530)

The word derives from the past participle of the Middle English *sacre,* to consecrate; from the Old French *sacre*; and from the Latin *sacrare,* sacred (*The American Heritage Dictionary of the English Language* 2000, p. 1530).

The use of sand is both long reaching through time and embued with a sacred nature. As an example, the Navajo people made sand paintings for use in ritual healing ceremonies (Markell 2002, p. 38; Weinrib 1983). Not only have the counseling setting and space been characterized as sacred (Cashwell et al. 2007, p. 69) but sandplay, and Kalffian sandplay in particular, have been described as establishing sacred space (Booth and O'Brien 2008, p. 87).

It is difficult as a therapist practicing sandplay to divorce oneself from the sacred; sandplay has its roots in understandings of the sacred. Jungian psychology, one of the roots of sandplay, is inextricably tied to a search for and an awareness of the spiritual and the sacred. Jung was interested not in

> the treatment of neurosis, but with the approach to the numinous. But the fact is that the approach to the numinous is the real therapy, and inasmuch as you attain to the numinous experience you are released from the curse of psychology. Even the very disease takes on a numinous character. (Jung 1973, p. 377)

Jung derived this concept of the numinous from Otto, who coined the term in 1958 from the Latin *numen,* meaning deity. Otto (1958, p. 12), a German theologian, suggested that the world and human life were created by God and, therefore, the awe of human beings is the cause of the existence of God (Shi 2005, p. 116). Otto described the experience of the numinous as

> sweeping like gentle tide, pervading the mind with tranquil mood of deepest worship. It may pass over into a more set and lasting attitude of the soul, continuing as it were, thrillingly vibrant and resonant, until at last it dies away and the soul resumes its "profane," non-religious mood of everyday experience.

The experience of the numinous has been credited with serving as the basis for humanity's experience of God or a higher power (Corbett 1996, p. 8).

Weinrib (1983, p. 163) defined the quality of the numinous in sandplay as "the divine force or potency ascribed to objects or beings regarded with awe." She described how Dora Kalff came to recognize sandplay's potential as a means by which clients might access the numinous:

> In her practice she saw that in the sand something activated by the mind brought forth a concrete creation which in the intuitive way of women brought forth insight, wisdom and the *numinous* experience. Relativization of the ego via encounter with the Self was experienced as numinous and was expressed in unmistakably religious symbols. (Weinrib 1983, p. 40. Emphasis in original)

Jung (1921, p. 243, quoted in Dourley 2006, p. 48) believed that "God's action springs from one's own inner being." Indeed, Jung appears to have identified the Self with the symbols of God (Heisig 1999, p. 95). According to Jung, the ego represents the center of consciousness and the Self represents psychological integrity or wholeness; this includes both the conscious and the unconscious (Jung 1928). Jung explained, "[T]he ego is not only the *centrum* of my field of consciousness, it is not identical with the totality of my psyche...the Self is the subject of my totality: hence it also includes the unconscious psyche" (Jung 1923, in Staub de Laszlo 1993, p. 309). He also stated, repeating the Gnostic saying, "God is an infinite circle (or sphere) whose center is everywhere and the circumference nowhere" (Jung 1961, p. 325).

Jung postulated, as well, that "[t]he symbols of divinity coincide with those of the Self: what, on the one side, appears as a psychological experience signifying psychic wholeness, expresses on the other side the idea of God" (Jung 1961, p. 339, 1961/2002, p. 10). Awareness of Self may, indeed, be a kind of sacred experience that produces awe (Liang 2012, p. 753).

Jung characterized his task as the "cure of souls" (Jung 1969, p. 353). He did not "attribute a religious function to the soul." Rather, he argued that the soul is *naturaliter religiosa* (Jung 1944/1999, p. 190), meaning that the soul is religious by nature. He defined the soul as the heart or the essence of a person that is endowed with the potential for a conscious relationship with the deity (Jung 1944/1999, p. 1987, par. 9–11). A psychoneurosis, he said, "must be understood, ultimately, as the suffering of a soul which has not discovered its meaning …." The cause of that suffering, he asserted, can only be "spiritual stagnation or psychic sterility" (Jung 1938, par. 497, excerpted in Brutsche 1999, p. 279).

Various aspects of sandplay also derive from Zen Buddhism, which had fascinated Dora Kalff. Buddhism is both a religion and a way of life, comprising numerous rituals and practices designed to assist its followers in their efforts to attain the ultimate goal. That goal varies depending upon the particular school of Buddhism, but is generally pre- or postmortal nirvana (Schumann 1973, Table VII). The specific practices to be utilized similarly vary across schools but may include meditation, repeated recitation of mantras, reflection on koans, and/or alignment of one's life activities and perspective to encompass and reflect delineated principles (Schumann 1973).

In all schools of Buddhism, the Buddha is deemed worthy of respect as a wise teacher and a guide. Some schools of Buddhism believe that the Buddha is a heavenly being, a deity, a savior, or a transcendent being (Schumann 1973, pp. 98–99).

Zen Buddhism challenges the individual to look inside of him- or herself. D.T. Suzuki, from whom Dora Kalff learned about Zen Buddhism, noted,

> Zen abhors words and concepts and reasoning based on them…we generally make too much of ideas and words thinking them to be facts themselves …. Zen upholds as every true religion does direct experience of Reality …. (Suzuki 1938, pp. 48–49)

Quoting from the Gandavyûha-Sûtra, Suzuki continued,

> Self-realisation never comes from mere listening and thinking …. By merely listening to it, thinking of it, and intellectually understanding it, you will never come to the realization of any truth …. The ultimate truth is a state of inner experience…and as it is beyond the realm of words and discriminations it cannot be adequately expressed by them. (Suzuki 1938, pp. 49–50)

In a sense, the Buddhist journey toward self-understanding can be analogized to Jung's search for individuation. Sandplay is similar; it is the client's vehicle in his or her search for Self. As Martin Kalff (2000, p. viii) explained, sandplay's

> emphasis [is] on creating a space that awakens and supports the self-making strengths of the patients. This quality of and play resembles an important aspect of Zen, in which the person is thrown back upon him or herself. Both sandplay and Zen emphasize that realization cannot be found in outer authorities, such as teachers or writings, but ultimately only within oneself.

4.3.1 Sacred Space and Process

As a therapist using sandplay, I also see sacredness in the client's repeated return to the sand tray, and the process by which this occurs. The desire of humans to connect to the sacred is evident from and throughout human history (Warfield 2012, p. 1). A physical structure or site may come to be regarded as a point through which individuals may connect with the sacred or divine (Warfield 2012, p. 1). Such a site may be identified through a pilgrimage, "a journey to a special place, in which both the journey and the destination have spiritual significance for the journeyer" (Davidson and Gitlitz 2002, p. xvii). The sacred space need not be explicitly religious or spiritual. Rather, it may have as its purpose a rite of passage, the fulfillment of a vow, the reaffirmation or discovery of identity, the atonement for transgression, and healing (Davidson and Gitlitz 2002). Relph (1976, p. 38) explained how such places come to be regarded as sacred:

> The places to which we are most attached are literally fields of care, settings in which we have had a multiplicity of experiences and which call forth an entire complex of affections and responses. But to care for a place involves more than having a concern for it that is based on certain past experiences and future expectations—there is also a real responsibility and respect for the place both for itself and what it is to yourself and to others. There is, in fact, a complete commitment to that place, a commitment that is as profound as any that a person can make ….

In a concrete sense, the sand tray becomes what Dora Kalff called "a free and protected space." It serves as the meeting place of four elemental powers—air, fire, water, and earth—and three cosmic regions—heaven, earth, and the underworld (Markell 2002, p. 59). On a psychic level, the sand tray as a sacred space serves as an anchor, reminding the client both of where he or she came from and what he or she once was (cf. Cooper-Marcus 1992, p. 89). The sand tray may become invested with spiritual meaning because it helps the client connect with past experiences, memories, and people (Csikzentmihalyi and Richberg-Halton, 1981). In essence, the sand tray itself provides "a symbolic lifeline to a continuous sense of identity" (Hummon 1989, p. 219). The client's repeated return to this free and protected space promotes recognition of the space itself as a "sacred and ritualized space-time dimension" (Markell 2002, p. 37). Steinhardt (2008, p. 200) has recognized this sacredness:

> There are some sandplays that I leave intact, if possible, for a few hours or more, until the felt presence of numinosity has evaporated and the objects and san can return to their unintentional states, until they are engaged in a new interaction.

The use of sand facilitates the process of self-discovery and self-knowledge. As Buber observed, "The origin of the world is dust, and man has been placed in it that he may raise the dust to spirit (Buber 1947, p. 29). Spiritual identity and knowledge of oneself can be nurtured in spaces that are conducive to meditation and that assist the individual in his or her efforts to understand his/her place in the world, the reason for his/her existence, and his or her purpose and role (Swan 1988). Moving one's hands through the sand may help a client connect with his or her feelings, facilitating a new awareness of oneself and one's relations to others.

It is not unusual for an individual to experience feelings of loss and grief at the dismantling of a sacred space (Mazumdar and Mazumdar 1993, p. 238). We see this historically with the destruction of the temple in the Old Testament. In the context of sandplay therapy, we do not dismantle the sand tray in front of our client. Rather, we remove the figures from the tray and smooth the sand, leaving a fresh palette, only after the client has left, the client having first had an opportunity to integrate the image of his or her completed tray.

I think of the sandplay process as a pilgrimage, a journey to one's inner depths, allowing the individual to connect with his or her authentic self, to others, and to the larger universe. A pilgrimage begins when an individual resolves to make a meaningful journey to a sacred place (Warfield 2012, p. 3). Subsequent steps involve making preparations, journeying to that place, the experience at the site, completion of the journey, and the individual's return home (Warfield 2012, p. 4). A client engaging in sandplay is, in fact, resolving to embark on a meaningful journey, one that requires repeated logistical preparations, such as time, money and transportation, and one that requires great fortitude and courage, for the client cannot know ahead of time what he or she will encounter along the way. Indeed, others have analogized the sandplay process to a pilgrimage:

> I like to compare the sandplay process with a pilgrimage. The pilgrim called client wanders from chapel to chapel. When coming to a chapel they meet a person, a therapist, who is with them in an attentive way. They then create a sand picture, a half conscious and half unconscious image (Amman 2008, p. 110).

When the pilgrimage is not of religious nature, it often encompasses a transformative process that addresses experiences of loss and suffering and transformation from illness to health (Winkelman and Dubisch 2005, p. xv). The transformation need not involve physical illness to health, but may center instead on a psychic transformation. The healing dynamics of pilgrimage include social, physical, symbolic effects, personal empowerment, and integration of self within collective (Winkelman and Dubisch 2005, p. xv).

During this pilgrimage, indeed a major component of this pilgrimage revolves around the client's encounter with the therapist and the space between them that arises from but is not synonymous with their relationship, a space that they cocreate together. The critical elements of a therapeutic relationship are acceptance, empathy, and congruence. These components can be conceptualized in sacred terms as love, compassion, and authenticity. In general, the therapeutic relationship may often encompass spirituality, and therapists may perceive themselves as participating in sacred activity (Leijssen 2008, p. 221). Rogers said of the therapeutic process that

> Our experiences in therapy involve the transcendent, the indescribable, the spiritual. I am compelled to believe that I, like many others, have underestimated the importance of this mystical, spiritual dimension. (Rogers 1980, p. 130)

If sandplay is sacred, and if sandplay therapists see themselves as participating in a sacred activity, does this suggest that the therapist is akin to God? Certainly, the therapist–client relationship has been analogized to the relationships between the client and God because of the need for trust, faith, and hope (Hutch 1983, p. 14).

Jung understood that the therapist could not labor under such a delusion. Rather, he cautioned that treatment is "an individual dialectical process, in which the doctor, as a person, participates just as much as the patient" and, because each individual and each individual's situation is unique, "the analyst must go on learning endlessly" (Jung 1951, p. 116).

Buber (1958, p. 21, 38) observed, "If I face a human being as my *Thou,* and say the primary word *I-Thou* to him, he is not a thing among things." The interaction between individuals in this manner has been conceived of as a meeting of the divine being viewed through that person, a means by which to address God (Lines (2002, pp. 111–112). Carl Rogers, in dialogue with the theologian Paul Tillich, explained,

I feel at times when I'm really being helpful to a client of mine ... there is something approximating an I-Thou relationship between us, then I feel as though I am somehow in tune with the forces of the universe or that forces are operating through me in regard to this helping relationship (Rogers 1989, p. 74).

During the process of Kalffian sandplay, the therapist is both permitted and privileged to accompany the client on his or her pilgrimage as he/she seeks to heal wounds, resolve conflicts, and discover a new way of being. The sandplay process frequently culminates in the client's encounter with the numinous, which may be reflected in one of the client's final trays.

Much of the sandplay process may proceed in silence, creating space for and a connection with the client through our silence as therapists and witnesses. Silence can be important therapeutically (Rajski 2003, p. 181), in that it allows for the release of unconscious psychological material and creates a communion between the therapist and the client (Rajski 2003, pp. 184–185). The silence is also a kind of meditation or prayer and the dynamic "a kind of divine psycho-therapy" (Keating et al. 2007, p. 65), creating space for the client's sacred journey.

Zornberg (2009, p. xviii) observed that, although "[s]peech and silence are essential to each other," they do not constitute the basis of communication. In the kabbalistic thought of Isaac Luria, God withdrew his fullness in order to create space for the world. The first creative act, then, was the creation of silence: "silence itself breaks, interrupts, the continuous murmur of the Real..." (Žižek 1992, p. 154). Buber's words, spoken in a completely different context, appear to anticipate the intersection of and interplay between the sand and the silence: "The altar of earth is the altar of silence, which pleases God beyond all else" (Buber 1947, p. 25). Indeed,

[s]ilence is so powerful a language that it reaches the throne of the living God. Silence is His language, though secret, yet living and powerful. (Kowalska 1987, p. 886, 887)

According to Rabbi Nahman, human beings are able to imitate God and create their own worlds only if space exists between them (Zornberg 2009, p. xviii). It is this vacant space, this void or abyss as it were, that allows individuals to create, to know others and, ultimately, to know themselves. In this space,

God is absent. But He is present, in the void itself. His present absence brings to life the absent presence that [is] the basis of communication. Perhaps this is what exists between people, separating and connecting them. The Talmud calls God "the third who is between them." He knows the truth of their relationship, however they may distort it. In a sense, He *is* its truth (Zornberg 2009, p. xix).

It is the empty space created by sandplay, through the vacant and beckoning sand, through the silence, this protected space that facilitates the client's communication and understanding both with him- or herself and with the therapist. This emptiness, created by the sand and the silence, beckons the client to embark on a journey to move from his or her fragmented self to a more integrated authentic self. Kirkegaard's (1992, p. 275) observations on the dangers of "too much" underscore the importance of this vacant, empty space:

> [T]he art of communication at last becomes the art of taking away… something … from someone…. When a man has his mouth so full of food that for this reason he cannot eat … does giving him food consist in stuffing his mouth even more or, instead, in taking a little away so that he can eat?

Indeed, words may be dangerous, unspeakable, compelling the speaker and the listener to visit dark, ruinous, and ruined places, too terrifying to contemplate. A whisper, though, "is neither speech nor silence … [but] is shaped precisely as a message, an address to the other" (Zornberg 2009, p. xxv). Like a whisper, sandplay

> is both an intentional address to the other and an appeal to hear the internal noise of otherness in which the message is couched. Only after the public interpretation of meaning has tamed the words of their human excess can they be uttered without fear (Zornberg 2009, p. xxv).

In the Jewish tradition, the whisper—*lechisha*—is considered to be a medium of prayer. (Zornberg 2009, p. xxvii). Likened to a whisper of meaning, sandplay reduces the otherwise unfathomable wounds to a manageable intensity, hinting at the past, the present, and the future, serving as a margin between the inside and the outside. In this way, "we do not merely ask God to hear our call for help, but also beg him, who knows what is hidden, to hear the silent cry of the soul" (Buber 1947, p. 23). It is in this "third area" between the therapist and the client that the distinctions between inner and outer disappear and where the sacred dimension of the relationship and the journey can be sensed:

> We must move beyond the notion of life consisting of outer and inner experiences and enter a kind of "intermediate realm" that our culture has long lost sight of and in which the major portion of transformation occurs. As we perceive such a shared reality with another person, and as we actually focus on it, allowing it to have its own life, like a "third thing" in the relationship, something new can occur. The space that we occupy seems to change, and rather than being the subjects, observing this "third thing," we begin to feel we are inside it and moved by it. We become the object, and the space itself and its emotional states are the subject. In such experiences, the old forms of relationship die and transform. *It is as if we have become aware of a far larger presence in our relationship, indeed a sacred dimension.* (Schwartz-Salant 1998, pp. 5–6, emphasis added)

During the course of this pilgrimage, the sandplay client encounters not only sand but also sometimes actual fire and often water in addition to sand. Fire, often thought of as a destructive force, may represent instead a potential renewal, a connection between the material and the ephemeral, and the link between life and death (Bachelard 1964, p. 16). Fire has been conceived of as a deity and many religions believe that a divine spirit resides within each person, often assuming the form of a fire that can be either fanned or quenched (Ronnberg 2010, pp. 82–84).

Water may represent a vehicle for physical or spiritual cleansing, a source of life, or an instrument of regeneration and renewal (Chevalier and Gheerbrant 1994, p. 1081). Inhabitants of the Near East during the time of the Old Testament viewed water as a sign of blessings (Chevalier and Gheerbrant 1982, p. 1083). The New Testament compares the heart to a parched land, awaiting divine revelation, just as the land would await rain to quench its thirst (Deut. 32:2. See also Psalm 42:1).

What the individual brings to sandplay is also sacred. Although each individual seeks communication, it is also true that

> each individual is an isolate, permanently non-communicating, permanently unknown, in fact unfound …. At the center of each person is an incommunicado element, and this is sacred and most worthy of preservation. (Winnicott 1965, p. 185)

Sandplay facilitates a connection between the conscious and the unconscious:

> The unconscious awakens in the selection of figures and the shaping of the sandplay, and at the same time, the ambitions and purposeful qualities of the conscious mind are silenced. In this way, sandplay promotes what Jung referred to as the transcendent function, making possible a completely new outlook on life. (Kalff 2000, p. xi)

In the context of sandplay, the client's transformative and often numinous experience that has come about through his or her connection to the sacred space, the concrete symbolic images, and the underlying primordial images may be akin to a revelation, that is, a communication from God (Martinez 2011, p. 2).

4.4 The Clients Revisited

If I were to look at Dora's tray as a minister-therapist, my first inclination would be to focus on the possible meaning and import of the religious symbols that she has used. The Madonna, as the Virgin Mother of God, "symbolizes Earth directed heavenward and thus becoming Earth transfigured …. Earth of light" (Chevalier and Gheerbrant 1994, pp. 1070–1071). Could she have sensed on an unconscious level that her own adult child had been diagnosed with terminal cancer, although no one had disclosed that information to her? Perhaps Dora sought protection for her son from Mary herself, a mother who had seen her own Son suffer. Or, perhaps she hoped that Mary would carry to God for her, her prayer for the health and life of her son.

Angels serve as intermediaries between God's kingdom and the material world; they are his messengers, his ministers, and guardians. They are "warning signs of the divine presence" and may serve as "heralds or agents of divine intervention" (Chevalier and Gheerbrant 1994, p. 22). Might she have been hoping for divine intervention to save her son from death or, at the very least, protection and assistance during his illness and his passage onward?

Dora's brief foray into the sand may have enabled her to begin to acknowledge and process the possibility of her son's illness on an unconscious level, since it could not be voiced yet on a conscious level. How could she ask if he were ill when no one around her was willing to openly raise the issue? But the tray may also have provided an opportunity for Dora to garner spiritual support in her struggle with depression and her unknowing acknowledgement of her son's illness. As agents of divine intervention, the angels may have represented a prayer to God for her son. Perhaps, as a child of God herself, the angels may have carried her prayer for the cessation of her own suffering to the ears of her Lord.

But, if I step back and away from my own focus on the sense of the spiritual that seems to be present in both the tray and in the space between us, perhaps there may be an entirely different significance to Dora's tray. Does she see herself as the angel, or the Madonna–mother or both, there to protect her children? Ultimately, we can never know for certain why Dora felt drawn to the angel or the Madonna, but it is critical that all such possibilities be considered.

We have unanswered questions with Dustin, as well. We cannot know with certainty why Dustin chose the Buddha figure or what it may have represented to him. Perhaps it represented his search for balance or for a glimpse of the divine elements of life on this earth or a hope that he would find compassion in the sandplay process, both from the therapist and from himself. Or, perhaps this choice of symbol reflected transference, a projection onto the therapist of an assumed wisdom. Or, perhaps it had nothing to do with the sacred, but represented for Dustin, instead, a small happy figure, such as he might have been as a child prior to his sexual abuse.

We can also look at Dustin's tray spatially. It is as if the different major scenes of his life, complete with characters, are played out in diagonal waves in the tray. We have the wave on the left of the drunk grandfather, the wave of the abuse, the wave of Dustin with his grandmother, the wave to the farthest right of his mother, and her other children. Perhaps the schoolhouse and the Buddha, located so close to the wave of the abuse, represent the place in Dustin's head where he went at the time of the abuse, a dissociation to a calmer, sweeter, more balanced, and centered place that he now seeks once again.

Both of these cases illustrate the beauty and the danger inherent in sandplay. As sandplay therapists, we recognize that sandplay allows and, indeed, facilitates connection to the numinous. We must take care, however, to avoid allowing our knowledge of the process, cloud the client's process. We can never know, truly know, why a client has chosen a particular symbol or placed that symbol in a particular position, perhaps in relation to others, even as we see the client's process unfolding. The silence of sandplay, whether sacred or not, creates the space in which this process can unfold. If I were to interject into this process my own sense and feeling of what is there, I could inadvertently misdirect the client to a place that is neither important nor relevant for the client, that moves the client from the place where he or she needs to be at that time to accomplish what he or she then needs or, at worst, harms the client. Dora, for example, may not have surmised on any level that her son was dying. Indeed, perhaps the psychiatrist and psychologist were correct and if I had attempted to address this issue with her, even indirectly, it would have led to a deepening of her depression. Supervision allowed me to become better aware of my own sense of awe and enhanced my ability to see the clients and their experience of their own processes.

4.5   Conclusion

This chapter focused on the issues of co-transference that may arise in working with sandplay clients whose trays evidence spiritual elements and potential associated ethical issues. While often understood as a clinical issue to be explored in supervision, co-transference that remains unaddressed or is unaddressed inappropriately may constitute an ethical issue related to practice competence and the failure of the therapist to take reasonable steps to avoid harming the client. The specific harm that may result depends upon the client's underlying mental state and other salient aspects of his or her life. As we have seen from the case of Dora, this may include an exacerbation of depression. Other potential harm might result, again depending upon the client's particular circumstances, e.g., exacerbation of religiously oriented delusions.

References

American Psychological Association. (2010). *Ethical principles of psychologists and code of conduct.* Washington, D.C.: American Psychological Association. http://www.apa.org/ethics/code/. Accessed 6 February 2015.

Amman, R. (2008). Supervision in an international, multilingual, and multicultural therapy world. In H.S. Friedman & R. Mitchell Rogers (Eds.), *Supervision of sandplay therapy* (pp. 107–112). New York: Routledge.

Bachelard, G. (1964). *The psychoanalysis of fire* (trans: A.C.M. Ross). Boston: Beacon.

Barker, R. L. (2003). *The social work dictionary*, 5th ed. Washington DC: NASW.

Booth, R., & O'Brien, P. J. (2008). An holistic approach for counselors: Embracing multiple intelligences. *International Journal of Advances in Counselling, 30,* 79–92.

Bradway, K. (1991). Transference and countertransference in sandplay therapy. *Journal of Sandplay Therapy, 1*(1), 25–43.

Bradway, K., & McCoard, B. (1997). *Sandplay: Silent workshop of the psyche.* New York: Routledge.

Buber, M. (1947). *Ten rungs: Collected hasidic sayings* (trans: O. Marx). London: Routledge.

Buber, M. (1958). *I and thou* (trans: R.G. Smith). Edinburgh: T & T Clarke.

Cashwell, C. S., Bentley, D. P., & Bigbee, A. (2007). Spirituality and counselor wellness. *Journal of Humanistic Counseling, Education, and Development, 46,* 66–81.

Chevalier, J., & Gheerbrant, A. (1982). *The penguin dictionary of symbols.* New York: Penguin.

Chevalier, J., & Gheerbrant, A. (1994). *The penguin dictionary of symbols.* London: Penguin.

Cooper Marcus, C. (1992). Environmental memories. In I. Altman & M. Low (Eds.), *Place attachment* (pp. 87–112). New York: Plenum.

Corbett, L. (1996). *The religious function of the psyche.* London: Routledge.

Csikzentmihalyi, M., & Richberg-Halton, E. (1981). *The meaning of things: Domestic symbols and the self.* New York: Cambridge University Press.

Davidson, L. K., & Gitlitz, D. M. (2002). *Pilgrimage from Ganges to Graceland: An encyclopedia.* Santa Barbara: ABC-CLIO.

Dourley, J. (2006). Jung and the recall of the gods. *Journal of Jungian Theory and Practice, 8*(1), 43–53.

Friedman, H. S, & Mitchell Rogers, R. (Eds.). (2008). Introduction. Supervision of sandplay therapy (pp. 1–10). New York: Routledge.

Heisig, J. W. (1999). Jung, christianity, and buddhism. *Nanzan Bulletin, 23,* 74–104. http://www. thezensite.com/non_Zen/Jung_Christianity_and_Buddhism.pdf. Accessed 20 Jun 2013.

Hummon, D. M. (1989). House, home and identity in contemporary American culture. In S. M. Low & E. Chambers (Eds.), *Housing culture and design: A comparative perspective* (pp. 207–228). Philadelphia: University of Pennsylvania Press.

Hutch, R. A. (1983). An essay on psychotherapy and religion. *Journal of Religion and Health, 22*(1), 7–18.

Jung, C. G. (1921). Psychological types. In Collected works of C.G. Jung, vol. 6. Quoted in J. Dourley. (2006). Jung and the recall of the Gods. *Journal of Jungian Theory and Practice, 8*(1), 43–53.

Jung, C. G. (1923). Psychological types, or the psychology of individuation (trans: H.G. Baynes). In V. Staub de Laszlo (Ed.), *The basic writings of C.G. Jung* (pp. 230–357). New York: Random House, Inc.

Jung, C. G. (1928). *Self and unconsciousness.* (trans: R.F.C. Hull). Bollingen Series XX. Princeton: Princeton University Press.

Jung, C. G. (1938). Psychology and religion. Collected works of C.G. Jung, vol. 11, par. 497. Excerpted in P. Brutsche. (1999). Illness and creativity. In M. A. Mattoon (Ed.), Destruction and creation: Personal and cultural transformations. Proceedings of the fourteenth international congress for analytical psychology (pp. 279–290). Einsiedeln: Daimon Verlag.

Jung, C. G. (1944/1999). Introduction to the religious and psychological problems of alchemy. In H. Stein (Ed.), *Jung on Christianity* (pp. 181–212). Princeton: Princeton University Press.

Jung, C. G. (1951). Fundamental questions of psychotherapy. In H. Read, M. Fordham, G. Adler, & W. McGuire (Eds.), *Collected works of C.G. Jung, vol. 16* (trans: R.F.C. Hull) (pp. 111–125). Princeton: Princeton University Press.

Jung, C. G. (1961). *Collected works of C.G. Jung, vol. 9, part 1.* New York: Pantheon.

Jung, C. G. (1961/2002). *Flying saucers: A modern myth of things seen in the skies.* New York: Routledge.

Jung, C. G. (1969). *Psychoanalysis and the cure of souls.* In Collected works of C.G. Jung, Vol. 11. 2nd ed. (pp. 348–354). Princeton: Princeton University Press.

Jung, C. G. (1973). *C.G. Jung letters, vol. I.* G. Adler with A. Jaffé (Eds.). Princeton: Princeton University Press.

Kalff, M. (2000). Forward. In D. M. Kalff (Ed.), *Sandplay: A psychotherapeutic approach to the psyche* (pp. v–xv). Cloverdale: Temenos.

Keating, T., Pennington, B., & Clarke, T. (2007). *Finding grace at the centre, 3rd ed.* Woodstock: Skylight Paths.

Kirkegaard, S. (1992). *Concluding unscientific postscript in the philosophical fragments: A mimic-pathetic-dialectic composition* (trans: H.V. Hong & E.H. Hong). Princeton: Princeton University Press.

Kowalska, G. (1987). *The diary: Mercy in my soul.* Stockbridge: Marian.

Leijssen, M. (2008). Encountering the sacred: Person-centered therapy as a spiritual practice. *Person-centered and Experiential Psychotherapies, 7*(3), 218–225.

Liang, H. (2012). Jung and Chinese religions: Buddhism and Taoism. *Pastoral Psychology, 61,* 747–758.

Lines, D. (2002). Counseling within a new spiritual paradigm. *Journal of Humanistic Psychology, 42*(3), 102–123.

Loue. S. (2009). A prologue to sandplay with an inner-city gay man: A case study. *Journal of Sandplay Therapy, 18*(1), 107–116.

Markell, M.J. (2002). *Sand, water, silence—The embodiment of spirit.* London: Jessica Kingsley.

Martinez, I. (2011). Reading for psyche: Numinosity. *Journal of Jungian Scholarly Studies, 7*(4), 1–15.

Mazumdar, S., & Mazumdar, S. (1993). Sacred space and place attachment. *Journal of Environmental Psychology, 18,* 231–242.

National Association of Social Workers. (2008). *Code of ethics.* http://www.socialworkers.org/pubs/code/code.asp. Accessed 6 February 2015.

Otto, R. (1958). *The idea of the holy* (trans: J. W. Harvey). New York: Oxford University Press.

Racker, H. (1968). *Transference and countertransference*. New York: International Universities Press.

Rajski, P. (2003). Finding god in the silence: Contemplative prayer and therapy. *Journal of Religion and Health, 42*(3), 181–190.

Relph, E. (1976). *Place and placelessness*. London: Pion.

Rogers, C. R. (1980). *A way of being*. Boston: Houghton Mifflin.

Rogers, C. R. (1989). A newer psychotherapy. In H. Kirschenbaum & V. L. Henderson (Eds.), *The Carl Rogers reader* (pp. 63–76). Boston: Houghton Mifflin.

Ronnberg, A. (Ed.). (2010)*The book of symbols*. Cologne: Taschen.

Schumann, H. W. (1973). *Buddhism: An outline of its teachings and schools* (trans: G. Feuerstein). Wheaton: Quest Books.

Schwartz-Salant, N. (1998). *The mystery of human relationship: Alchemy and the transformation of self*. New York: Routledge.

Shi, D. L. (2005). *Exploration on psyche: Mysterious archetype*. Harbin: Heilongjiang.

Smith, H. (1991). *The world's religions*. San Francisco: HarperSanFrancisco.

Staub de Laszlo, V. (Ed.). (1993). *The basic writings of C.G. Jung*. New York: Random House, Inc.

Steinhardt, L.F. (2008). Supervision in sandplay: The art therapist as sandplay supervisor. In H. S. Friedman & R. Mitchell Rogers. (Eds.). *Supervision of sandplay therapy* (pp. 198–214). New York: Routledge.

Suzuki, D. T. (1938). Zen Buddhism. *Monumenta Nipponica, 1*(1), 48–57.

Swan, J. (1988). Scared places in nature and transpersonal experiences. *ReVision, 16,* 21–26.

The American Heritage Dictionary of the English Language. (2000). *The American Heritage Dictionary of the English Language*, 4th ed. Boston: Houghton Mifflin.

Warfield, H.A. (2012). Quest for transformation: An exploration of pilgrimage in the counseling process. *VISTAS, 1*–5. http://www.counselingoutfitters.com/vistas/vistas12/Article_35.pdf. Accessed 20 Aug 2013.

Weinrib, E.L. (1983). *Images of the self: The sandplay therapy process*. Boston: Sigo.

Winkelman, M., & Dubisch, J. (2005). Introduction: The anthropology of pilgrimage. In J. Dubisch & M. Winkelman (Eds.), *Pilgrimage and healing* (pp. ix–xxxvi). Tucson: University of Arizona Press.

Winnicott, D. W. (1965). Communicating and not communicating leading to a study of certain opposites. In D. W. Winnicott (Ed.) *The maturational processes and the facilitating environment* (pp. 178–191). London: The Hogarth Press and the Institute of Psycho-analysis.

Žižek, S. (1992). *The parallax view*. Cambridge: Harvard University Press.

Zornberg, A. G. (2009). *The murmuring deep: Reflections on the Biblical unconscious*. New York: Schocken.

Chapter 5
Ethical Issues in Sandplay Research

Sana Loue

5.1 Introduction

There is a growing interest in and need for research that explicates the underlying mechanism of sandplay and assesses its efficacy and effectiveness. The vast majority of research relating to sandplay therapy to date has been conducted through case reports that follow a client over time. These case studies have often focused on clients' efforts to resolve specific issues, reduce problematic behaviors, or ameliorate depression (Ammann 1991; Kalff 1980; Weinrib 1983). Hong (2011) is one of the few sandplay therapists to have conducted outcomes research that utilized existing, validated instruments. Her study of the use of sandplay therapy with ten elementary school-level children assessed the therapeutic outcomes through the pre- and post-therapy administration of the Children Depression Inventory (Kovacs 1979), the Rorschach test (Exner 1986), the Teacher's Report Form (Achenbach 1991), and the House-Tree-Person Drawing. Hong herself noted the limitations of her research, which did not utilize a control group, did not follow children for as long a period of time as might be needed therapeutically and relied on a diverse pool of therapists, confounding the assessment of the therapy's impact, as distinct from the therapist's impact (Hong 2011, pp. 49–50).

Despite the growing interest in and expanding body of sandplay research, there has been little to no discussion relating to the ethical issues associated with research efforts that may be undertaken. This chapter represents an initial effort to address this gap.

S. Loue (✉)
School of Medicine, Case Western Reserve University, Cleveland, OH, USA
e-mail: sana.loue@case.edu

© Springer International Publishing Switzerland 2015
S. Loue (ed.), *Ethical Issues in Sandplay Therapy Practice and Research*,
SpringerBriefs in Social Work, DOI 10.1007/978-3-319-14118-3_5

5.2   The Ethical Framework for Research

The atrocities of the Nazi experiments on unwilling concentration camp prisoners—
e.g., injection of dye into eyes in an attempt to change eye color, implantation of
cow embryos into human women—led to the enunciation of principles designed to
govern research. These principles are embodied in the Nuremberg Code (1946) and
are set forth in Table 5.1 below.

It is not at all suggested here that research efforts relating to sandplay therapy
are in any way akin to the Nazi experiments. Indeed, any research conducted to assess the efficacy and effectiveness of sandplay is far different than the experiments
that gave rise to the Nuremberg Code. Nevertheless, basic ethical principles that
govern experimental research, such as clinical trials of drugs and devices or behavioral research to evaluate an intervention, are also applicable to psychotherapy
and counseling research (Etherington 2007). The British Psychological Society has
explained why principles are necessary to guide psychological research:

> Ethical research conduct is, in essence, the application of informed moral reasoning,
> founded on a set of moral principles …. By openly stating the values that underpin our pro-
> fession, at this historical point, we make them available for discussion and debate, as well
> as allowing the possibility of clarification and change. Moreover, locating the responsibility
> for developing adequate ethics protocols firmly and squarely with researchers themselves
> can be achieved by appealing to explicit, core principles at a sufficiently high level of
> abstraction that the likelihood of individual cases falling outside of them is minimal. (Brit-
> ish Psychological Society 2010, p. 7)

The provisions of the Nuremberg Code give rise to three basic principles: respect
for persons, beneficence, and justice. Respect for persons encompasses the concept
of autonomy and serves as the basis for the requirement that research with human
beings can be conducted only with the informed consent of the individual. How
we understand autonomy depends upon our notion of personhood. In the US context, this is often interpreted as reference to individual rights, self-determination,

Table 5.1 Provisions of the Nuremberg Code

Voluntary consent is essential
The experiment must yield fruitful results for the good of society
The experiment should be based on the results of animal experimentation and a knowledge of the natural history of the disease under study to justify performance of the study
The experiment should be conducted to avoid all unnecessary physical and mental suffering and injury
In general, no experiment should be conducted where there is a priori reason to believe that death or disabling injury will occur
Proper precautions must be taken to provide adequate facilities to protect the participant against the risk of injury, disability, or death
The experiment may be conducted by only scientifically qualified persons
The participant may end the experiment
The researcher must be prepared to end the experiment at any time

and privacy (De Craemer 1983). Beneficence refers to the researcher's obligation to maximize good to the research participants. This principle is sometime parsed into two, the second being nonmaleficence, or the obligation to minimize harm to the research participants. Justice, frequently interpreted as distributive justice, is predicated on the researcher's responsibility to equitable distribute the benefits and burdens of research across groups.

Similarly, the Helsinki Declaration, which has been revised on multiple occasions, reflects these principles while providing additional guidance on ethical considerations in conducting research. The Helsinki Declaration permits surrogate consent, by which parents and guardians can consent to a child's participation in research, recognizes that some groups may be especially vulnerable as research participants, and delineates the circumstances under which participation in research can be truly considered to be both informed and consensual.

Similar principles governing the ethical conduct of research have been recognized by mental health professional associations in various countries. The British Psychological Society advises that the ethical conduct of research requires the psychologist-researcher to: (1) recognize and respect the autonomy and dignity of persons, (2) conduct only research with scientific value, (3) recognize and enact social responsibility, and (4) maximize the benefits and minimize the harm to research participants (British Psychological Society 2010). Like the Helsinki Declaration, the British Psychological Society explicitly provides that parents may consent to their children's participation in research. The Canadian Psychological Association (2010) lists four principles underlying the ethical conduct of psychological research: (1) respect for the dignity of persons, (2) responsible caring, (3) integrity in relationships, and (4) responsibility to society. These principles roughly equate to those derived from the Nuremberg Code and encompass concepts of informed consent, protection of vulnerable persons, the minimization of risk, and the maximization of benefit. The principles delineated by the Psychological Society of Ireland (2011) for the ethical conduct of research by psychologists are similar and reflect comparable understandings: (1) respect for the rights and dignity of the person, (2) competence, (3) responsibility, and (4) integrity.

5.2.1   Respect for Persons

As indicated above, the principle of respect for persons encompasses the concept of autonomy. This principle suggests that (1) individuals and groups may be different in ways that are relevant to their worldview and their response to any variety of situations; (2) the researcher must respect these differences and fashion their research protocols in a way that is sensitive to these varying understandings, while still ensuring that fundamental principles of informed consent are observed; and (3) the researcher is responsible for ensuring that individuals with impaired or diminished autonomy who are participating in the research are protected from harm or abuse.

It has been suggested that a person can act autonomously only if he or she "acts (1) intentionally, (2) with understanding, and (3) without controlling influences"

(Faden and Beauchamp 1986, p. 238). In order to act with understanding, the individual must have the capacity to do so and must have the information necessary for understanding.

Capacity Capacity refers to the ability of an individual to evidence a choice, the ability to understand relevant information, the ability to appreciate a situation and its consequences, and the ability to manipulate information rationally. This is different from competence, which is a legal determination relating to an individual's ability to care for him- or herself and/or his or her financial affairs.

There is a presumption at the beginning of all research studies that a prospective adult participant has the capacity to consent unless there is reason to believe either that he or she does not have capacity or that the capacity to consent may be limited in some way. (Children are by law presumed to lack adequate capacity to consent although the age at which childhood ends and adulthood begins may differ across states in the USA and across countries.) Decision-making ability in the context of participation in research requires that the individual be able to understand basic study information, including the procedures to be performed, the risks associated with participation, the potential benefits he or she may gain from participation, alternatives to study participation, the difference between research interventions and established therapy, and the individual's ability to refuse to participate without suffering a penalty (cf. Dresser 2001).

Socioeconomic disadvantage, in particular, is believed to be "a critical concern in the context of behavioral health research" (De Vries et al. 2004). This stems from existing inequities and lack of access to income, housing, employment, and health care (National Bioethics Advisory Commission 2001).

Consider, for example, conducting research with heroin-dependent or heroin-addicted persons, a situation that illustrates how a number of ethical issues related to informed consent might arise, including issues relating to socioeconomic disadvantage.

Case Illustration
A therapist employed by a community-based mental health/substance use recovery facility wants to evaluate the effectiveness of sandplay therapy with individuals who are in recovery from heroin use. He/she will compare their progress toward abstinence and resumption of employment with that of individuals whose therapists utilize cognitive behavioral therapy. The therapist conducting the research must ensure that at the time the research is explained to prospective participants, they are fully lucid and are not experiencing the effects of heroin or any other substance. If they are not, the therapist may not enroll them in the study at that time. She may, however, schedule an appointment at a later time when the individual can return to discuss his or her potential participation in the research.

On the one hand, it could be argued that persons addicted to heroin can never be competent to consent to enrollment in a study because they are obsessed with the drug, they lack a stable set of values because of their addiction, whatever values they espouse are no longer truly theirs due to the impact of their addiction to heroin and, consequently, they cannot be accountable for any decision (Charland 2002).

This perspective, though, is problematic for several reasons. First, it presumes that all heroin-addicted individuals lack capacity to consent to participate in research despite the general presumption at the commencement of research studies that a prospective adult participant has capacity to consent (National Bioethics Advisory Commission 1998) and the nature of the proposed research. This position essentially equates an inability or unwillingness to say "no" to a lack of capacity (Ling 2002). This position also rests on a gross exaggeration of the impact of addiction; the fact of a diagnosis of addiction or drug dependence is relevant to, but not determinative of, the issue of capacity (Carter and Hall 2008). Finally, a determination that heroin-dependent persons could not be held responsible for their decisions and their conduct in the research context due to their drug dependence raises additional issues regarding their capacity in the clinical care and criminal law contexts.

In contrast to this unequivocal view of heroin addicts as lacking a stable set of values, another scholar has suggested that addicts are cognizant of the choices available to them, for example, participation in an unproven therapy to reduce or eliminate heroin use versus life on the streets supported through begging and criminal activity, and that they are able to assess the extent to which each such choice is consistent with their values to arrive at a decision (Perring 2002). The use of needle exchange programs, for instance, has established that injecting heroin-dependent persons are, despite their addiction, able to weigh the risks and benefits of using such a service in order to reduce potential health threats.

Nevertheless, heroin-dependent individuals are often considered vulnerable persons within the context of research. Vulnerable participants are those individuals with "insufficient power, prowess, intelligence, resources, strength or other needed attributes to protect their own interests through negotiations for informed consent" (Levine 1988). The ability of heroin addicts to protect their own interests may be temporarily diminished if they are undergoing the acute effects of heroin use or of withdrawal, but the use of the drug has not been found to affect attention or memory (Lundqvist 2005). Indeed, the capacity to provide informed consent may be understood as fluctuating, a phenomenon that has been recognized in considering the enrollment of mentally ill persons in research (National Bioethics Advisory Commission 1998).

Although heroin-addicted persons as a group may be disempowered due to poverty, imprisonment, and/or stigmatization, this status does not, however, negate their ability to participate. Rather,

> [c]onsiderable care should be exercised if a participant is unable to give personal consent on his or her own behalf for any reason, including lack of capacity due to … intoxication. (Bond 2004, p. 7)

Accordingly, therapist-researchers should develop protections to maximize the likelihood that prospective participants have capacity to consent at the time that they are solicited for their participation. A refusal to enroll any heroin-dependent person in sandplay therapy research based on his or her membership in the class of heroin-dependent persons, absent an individualized assessment of capacity to consent, would contravene the ethical principle of justice, discussed further below.

The issue of voluntariness may also arise in such research. As an example, some might argue that individuals who are addicted to a drug are incapable of giving consent voluntarily to participate in an intervention, such as an intervention study to assess the effectiveness of sandplay therapy, specifically because of their addiction and the hold that their addiction has over their behaviors. Others might assert that an individual with a heroin addiction is able to voluntarily choose participation in a study of an intervention that may assist him or her to end or reduce drug use; the alternative to reduction or cessation of drug use that faces heroin addicts is not whether to obtain heroin, but from whom (the dealer or the clinical trial) and at what cost (life on the streets, privacy, and personal freedom, risk of disease vs. supervision and loss of privacy).

Understanding and Information To act with understanding also suggests that the prospective research participant has been provided with adequate information regarding the nature of the research and its potential implications and consequences to enable him or her to make an informed choice regarding participation. Various international documents outlining the requirements for ethical biomedical and epidemiologic research delineate specific elements of information that must be provided to prospective research participants (Council of International Organizations for Medical Sciences 2002, 2009). Many of these elements are also included in US federal regulations that govern all research conducted in institutions that receive federal funding, for example, hospitals and institutions that receive Medicare or Medicaid payments and universities that receive federal research grants.

It might appear at first that these provisions have no relevance to sandplay therapy research since sandplay therapy research is not biomedical research such as a clinical trial for a new drug would be. However, a more in-depth examination of these informational elements suggests that "best practice" in sandplay therapy research would similarly recommend the inclusion of many of these requisite provisions of informed consent in order to reduce the possibility of harm to prospective research participants. Table 5.2 reframes the informational elements required by the Council of International Organizations for Medical Sciences for biomedical research so as to be relevant to sandplay research.

The American Psychological Association (APA) has delineated similar requirements for valid informed consent to participate in research:

(1) the purpose of the research, expected duration and procedures; (2) their right to decline to participate and to withdraw from the research once participation has begun; (3) the foreseeable consequences of declining or withdrawing; (4) reasonably foreseeable factors that may be expected to influence their willingness to participate such as potential risks, discomfort or adverse effects; (5) any prospective research benefits; (6) limits of confidentiality; (7) incentives for participation; and (8) whom to contact for questions about the research and research participants' rights. They provide opportunity for the prospective participants to ask questions and receive answers. (American Psychological Association 2014, par. 8.02(a))

Table 5.2 Informational elements required by the Council of International Organizations for Medical Sciences for informed consent to participate in research: suggested reframing for relevance to sandplay therapy research

The investigator must provide the following information, in a language or another form of communication that the individual can understand:
1. That the individual is invited to participate in research, the reasons for considering the individual suitable for the research, and that participation is voluntary
2. That the individual is free to refuse to participate and will be free to withdraw from the research at any time without penalty or loss of benefits to which he or she would otherwise be entitled
3. The purpose of the research, the procedures to be carried out by the investigator and the participant, and an explanation of how the research differs from routine therapy or counseling
4. For controlled trials, an explanation of features of the research design (e.g., randomization, double-blinding), and, in research where blinding is utilized, that the participant will not be told of the assigned treatment until the study has been completed and the blind has been broken
5. The expected duration of the individual's participation (including number and duration of visits to the research centre and the total time involved) and the possibility of early termination of the trial or of the individual's participation in it
6. Whether money or other forms of material goods will be provided in return for the individual's participation and, if so, the kind and amount
7. That, after the completion of the study, participants will be informed of the findings of the research in general, and individual participants will be informed of any finding that relates to their particular health status
8. That participants have the right of access to their data on demand, even if these data lack immediate clinical utility (unless the ethical review committee has approved temporary or permanent nondisclosure of data, in which case the participant should be informed of, and given, the reasons for such nondisclosure)
9. Any foreseeable risks, pain or discomfort, or inconvenience to the individual (or others) associated with participation in the research, including risks to the health or well-being of a participant's spouse or partner
10. The direct benefits, if any, expected to result to participants from participating in the research
11. The expected benefits of the research to the community or to society at large, or contributions to scientific knowledge
12. Whether, when and how any interventions proven by the research to be safe and effective will be made available to participants after they have completed their participation in the research, and whether they will be expected to pay for them
13. Any currently available alternative interventions or courses of treatment
14. The provisions that will be made to ensure respect for the privacy of participants and for the confidentiality of records in which participants are identified
15. The limits, legal or other, to the investigators' ability to safeguard confidentiality, and the possible consequences of breaches of confidentiality
16. The sponsors of the research, the institutional affiliation of the investigators, and the nature and sources of funding for the research
17. The possible research uses, direct or secondary, of the participant's counseling or other records

Table 5.2 (continued)

18. Whether the investigator is serving only as an investigator or as both investigator and the participant's therapist
19. The extent of the investigator's responsibility to provide counseling or therapy services to the participant
20. Whether treatment will be provided free of charge for specified types of research-related injury or for complications associated with the research, the nature and duration of such care, the name of the organization or individual that will provide the treatment, and whether there is any uncertainty regarding funding of such treatment
21. In what way, and by what organization, the participant or the participant's family or dependants will be compensated for disability or death resulting from such injury (or, when indicated, that there are no plans to provide such compensation)
22. Whether or not, in the country in which the prospective participant is invited to participate in research, the right to compensation is legally guaranteed
23. That an ethical review committee has approved or cleared the research protocol
24. in what way, and by what organization, the participant or the participant's family or dependants will be compensated for disability or death resulting from such injury (or, when indicated, that there are no plans to provide such compensation);
25. whether or not, in the country in which the prospective participant is invited to participate in research, the right to compensation is legally guaranteed;
26. that an ethical review committee has approved or cleared the research protocol

The APA additionally provides:

> Psychologists conducting intervention research involving the use of experimental treatments clarify to participants at the outset of the research (1) the experimental nature of the treatment; (2) the services that will or will not be available to the control group(s) if appropriate; (3) the means by which assignment to treatment and control groups will be made; (4) available treatment alternatives if an individual does not wish to participate in the research or wishes to withdraw once a study has begun; and (5) compensation for or monetary costs of participating including, if appropriate, whether reimbursement from the participant or a third-party payor will be sought. (American Psychological Association 2014, par. 8.02(b))

Social workers are also ethically obligated to ensure that participants in their research are adequately informed and protected. The Code of Ethics of the National Association of Social Workers (the USA) advises that:

> Social workers engaged in evaluation or research should obtain voluntary and written informed consent from participants, when appropriate, without any implied or actual deprivation or penalty for refusal to participate; without undue inducement to participate; and with due regard for participants' well-being, privacy, and dignity. Informed consent should include information about the nature, extent, and duration of the participation requested and disclosure of the risks and benefits of participation in the research.
>
> a. When evaluation or research participants are incapable of giving informed consent, social workers should provide an appropriate explanation to the participants, obtain the participants' assent to the extent they are able, and obtain written consent from an appropriate proxy.
> b. Social workers should never design or conduct evaluation or research that does not use consent procedures, such as certain forms of naturalistic observation and archival research, unless rigorous and responsible review of the research has found it to be

justified because of its prospective scientific, educational, or applied value and unless equally effective alternative procedures that do not involve waiver of consent are not feasible.

c. Social workers should inform participants of their right to withdraw from evaluation and research at any time without penalty.

d. Social workers should take appropriate steps to ensure that participants in evaluation and research have access to appropriate supportive services.

e. Social workers engaged in evaluation or research should protect participants from unwarranted physical or mental distress, harm, danger, or deprivation. (National Association of Social Workers 2008, par. 5.02(3)–(j))

In reviewing these elements, we must ask: What are the risks and benefits of participation in sandplay therapy? The British Psychological Society (2010) has indicated that any research characterized by one or more of the following presents greater than minimal risk to the participants:

- Research involving vulnerable groups
- Research related to sensitive topics, such as sexual behavior
- Research involving a significant element of deception
- Research that requires access to records containing information of a personal or confidential nature
- Research requiring access to sensitive information through a third party
- Research that may lead to increased psychological stress, anxiety, or humiliation
- Research involving invasive interventions not encountered in everyday life
- Research that may have an impact on the participant's employment or social standing
- Research that may lead to labeling by the researcher or participant, for example, the participant labels him or herself as "stupid"
- Research involving the collection of human tissue, blood, or other biological samples

It appears to this author that foremost among the risks that are potentially encountered in any of the above situations that are relevant to sandplay therapy research are those of loss of confidentiality and/or privacy and its consequences, retraumatization, and stigmatization.

Confidentiality and Privacy In general, mental health professionals conducting research are required ethically to maintain the confidentiality of the data that they collect from research participants. For example, the Code of Ethics of the National Association of Social Workers in the USA provides:

Social workers should respect clients' right to privacy. Social workers should not solicit private information from clients unless it is essential to providing services or conducting social work evaluation or research. Once private information is shared, standards of confidentiality apply. (National Association of Social Workers 2008, par. 1.07)

In the USA, the therapist-researcher's ability to assure confidentiality may be limited due to a duty to warn, state-imposed reporting requirements, and legal attempts to access the data. Similar legal limitations may exist in other countries; the ethical, if not the legal, obligation to warn is common across mental health professions across many countries. Although these issues may arise during any research, they may be especially likely to arise in studies conducted over an extended period of time.

It is beyond the scope of this chapter to review the duty to warn as it exists in multiple countries; for that reason, the discussion will focus on the obligation as it exists in the USA. In the USA, a "duty to warn" may exist as the result of a line of court cases that began in 1976 with the now-famous case of *Tarasoff v. Regents of the University of California*. The case involved a lawsuit by the Tarasoff family against the University of California and a psychologist at the Berkeley campus of the university for the death of their daughter Tatiana. Tatiana had refused the advances of another graduate student at Berkeley. The would-be suitor had revealed his intent to kill Tatiana during the course of counseling sessions with a psychologist at the school's counseling services. The psychologist and several colleagues sought to have this student involuntarily hospitalized for observation purposes, but he was released after a brief observation period, during which it was concluded that he was rational. He subsequently shot and killed Tatiana.

The majority of the court rejected the psychologist's claim that he could not have advised either the family or Tatiana of the threat because to do so would have breached the traditionally protected relationship between the therapist and the patient. Instead, the court held that when a patient "presents a serious danger ... to another [person], [the therapist] incurs an obligation to use reasonable care to protect the intended victim against such danger." That obligation could be satisfied by warning the intended victim of the potential danger, by notifying authorities, or by taking "whatever other steps are reasonably necessary under the circumstances" (*Tarasoff v. Regents of the University of California* 1976). The court specifically noted that the therapist–patient privilege was not absolute:

> We recognize the public interest in supporting effective treatment of mental illness and in protecting the rights of patients to privacy and the consequent public importance of safeguarding the confidential character of psychotherapeutic communication. Against this interest, however, we must weigh the public interest in safety from violent assault .... We conclude that the public policy favoring protection of the confidential character of patient-psychotherapist communications must yield to the extent to which disclosure is essential to avert danger to others. The protective privilege ends where the public peril begins.

Some later cases have followed the reasoning of the *Tarasoff* court. A New Jersey court ruled in *McIntosh v. Milano* (1979) that the doctor–patient privilege protecting confidentiality is not absolute, but is limited by the public interest of the patient. In reaching this conclusion, the court relied on the 1953 case of *Earle v. Kuklo* (1953), in which the court had stated that "a physician has a duty to warn third persons against possible exposure to contagious or infectious diseases." A Michigan appeals court held in *Davis v. Lhim* (1983) that a therapist has an obligation to use reasonable care whenever there is a person who is foreseeably endangered by his or her patient. The danger would be deemed to be foreseeable if the therapist knew or should have known, based on a professional standard of care, of the potential harm.

Courts are divided, however, on whether the patient must make threats about a specific, intended victim to trigger the duty to warn. The court in *Thompson v. County of Alameda* (1980) found no duty to warn in the absence of an identifiable victim. Another court, though, held that the duty to warn exists even in the absence of specific threats concerning specific individuals, if the patient's previous history suggests that he or she would be likely to direct violence against a person (*Jablonski v. United States* 1983).

Depending on the particular state, however, researchers may also be required to report instances of child sexual abuse, child abuse or neglect, elder abuse, or intimate partner violence that may be committed by or perpetrated on a research participant. Whether such an obligation exists often depends on the age and state of residence of the victim, the state's definition of the offense, the recency of the event, and the status of the reporter, that is, whether a researcher under that state's laws is a mandated reporter.

Confidentiality may also be limited due to a subpoena. A subpoena is an order from a court or administrative body to compel the appearance of a witness or the production of specified document or records. This discussion focuses on subpoenas issued to compel the production of records or documents associated with the research.

A subpoena can be issued by a court or administrative body at the state or federal level. The information sought may be believed to be important to the conduct of an investigation, a criminal prosecution, or a civil lawsuit. The issuance of subpoenas against researchers had become increasingly common (Auriti 2013) and they have been used as a mechanism to obtain data relating to identifiable research participants (e.g., Hayes 2011).

Certificates of confidentiality, available in some circumstances in the USA for research conducted within the USA, may potentially limit the extent to which research data may be obtained by subpoena. Certificates of confidentiality are issued by the appropriate institute of the National Institute of Health and other agencies of the US Department of Health and Human Services. Authority for their issuance derives from the section 301(d) of the Public Health Service Act, which provides that:

> The Secretary may authorize persons engaged in biomedical, behavioral, clinical, or other research (including research on mental health, including research on the use and effect of alcohol and other psychoactive drugs) to protect the privacy of individuals who are the subject of such research by withholding from all persons not connected with the conduct of such research the names or other identifying characteristics of such individuals. Persons so authorized to protect the privacy of such individuals may not be compelled in any Federal, State, or local civil, criminal, administrative, legislative, or other proceedings to identify such individuals.

Certificates are potentially available for research where the participants may be involved in litigation that relates to the exposure under study, such as occupational exposure to HIV; that collects genetic information; that collects data pertaining to participants' psychological well-being, their sexual attitudes, preferences, or practices or their substance use or other illegal activities or behaviors. A certificate of confidentiality is available only for research data collected in the USA; it is not available, for example, if a sandplay therapist in the USA (or elsewhere) is conducting the research outside of the USA. Additional details relating to certificates are available from the various websites sponsored by the Office of Extramural Research of the National Institutes of Health (http://grants.nih.gov/grants/policy/coc/appl_extramural.htm; http://grants.nih.gov/grants/policy/coc/back ground.htm; http://grants. nih.gov/grants/policy/coc/faqs.htm).

The validity of these certificates was once upheld by a New York court (*People v. Newman* 1973). However, their validity is subject to question because, in essence, they allow an agency of the federal government to limit the ability of the states to investigate and prosecute possible criminal activity and the ability of the courts and litigants in civil cases to obtain evidence that may be critical.

While a certificate of confidentiality may relieve the sandplay therapist-researcher of the legal duty to disclose specific information, it does not relieve him or her of

---

Case Illustration

A sandplay therapist wishes to conduct research using a series of case studies of individual clients. He is particularly interested in devising a method that tracks the clients' use of symbols with their progress toward resolution of their presenting issue. Specifically, he wishes to see if his male clients who are struggling with anger control issues and have a conviction for child sexual abuse use fewer "aggressive" figures, such as monsters and warring soldiers as they become better able to deal with and process anger. The informed consent process with the clients must provide the clients with information detailing the risks and benefits of participation. One potential risk in all such situations is an inadvertent breach of confidentiality. For these particular clients, it might have serious consequences if, for example, the therapist-researcher's data are accessed by way of a subpoena and law enforcement authorities were to decide that a client is unable to channel his aggression and insecurities in a manner that does not potentially endanger children. In some cases, depending upon the specifics of the situation and if the data were collected in the USA, the therapist-researcher may be able to obtain a certificate of confidentiality prior to initiating the study in order to protect the client data that will be used in the research from being accessed via subpoena. However, the certificate of confidentiality would not protect the information provided to the therapist outside of the research, e.g., clinical notes that existed prior to the initiation of the study and award of the certificate.

---

any ethical responsibility to do so. For example, a certificate of confidentiality may relieve the therapist-researcher of the obligation to report to designated authorities that a client-research participant is the current victim of elder abuse. It does not, however, relieve the therapist-researcher of any associated ethical obligation. The client-research participant must be fully informed as part of the informed consent process regarding the extent of confidentiality protection and what the therapist-researcher will report.

Voluntariness What constitutes a "controlling influence" varies across cultures. As an example, many Americans conceive of themselves as independent agents free to make decisions without consideration of or reference to either the opinions of others or the potential impact of their decisions on others. In contrast, individual identity in other cultures may rest on the idea of an "enlarged self;" individuals in these cultures see themselves not as autonomous agents, but as the aggregation and integration of various roles and relationships, each with corresponding responsibilities. To an individual raised at the altar of Western individualism, reference to and consideration of others' viewpoints may be interpreted as a "controlling influence." Nevertheless, where this consultation by a prospective research participant with others is voluntary, it is entirely consistent with the principle of respect for persons.

How is this relevant to sandplay research?

We can consider a somewhat typical situation. A client who presents for sandplay therapy signs an informed consent form for treatment. Maybe the form includes a paragraph specifying that, after an appropriate passage of time, the therapist may present the client's case at a conference or in a journal article. Or, maybe the therapist provides the client with a separate release form allowing him or her to use the client's case in this way. A client raised in the tradition of Western individualism may not give signing a second thought. In contrast, clients whose cultures embrace the concept of an extended self may be reluctant to agree without first considering the potential implications of their agreement on family members or perhaps even consulting with others. They may feel, for example, that public attention to their situation, even when their identity is masked, may somehow bring shame to the family or indicate disloyalty.

There is also the issue of the power differential that exists between the therapist-researcher and the client. In some cases, a researcher has no relationship with the individual research participant, such as when the researcher sends out a survey to everyone living in a specific neighborhood to determine whether the presence or absence of sidewalks affects individuals' ability to exercise. This is not, however, the case if a sandplay therapist wishes to conduct research using his or her own cases. As noted in a publication of the British Association for Counselling and Psychotherapy,

> Seeking consent for participation in a research study at some point after the person has entered counseling or psychotherapy (for example, contacting clients who are in the process of receiving therapy to invite them to take part in a follow-up interview) raises serious issues about the potential for coercion .... The dual relationship created by practitioners undertaking research on their own counseling or psychotherapeutic service is very likely to affect, either positively or negatively, both the therapy and the research. (Bond 2004, p. 7, 9)

An individual who is then obtaining therapy might feel that the sandplay therapist will not provide the same quality of care, will not listen as well, or will terminate services prematurely if the client does not agree to be part of a study. Even if the client has terminated with the sandplay therapist, he or she might fear that a refusal to participate in the research would lead to a refusal by the therapist to provide future services if the client wished to have them.

Various strategies can and should be utilized in an effort to minimize any potential risk to the client-research participant. The British Association for Counselling and Psychotherapy has recommended the following:

1. Care is taken to ensure that the undertaking of any research by the practitioner is both beneficial to the client and also consistent with the integrity of the research.
2. Thorough consultation, with both a research consultant or ethics committee, and the practitioner's counselling or psychotherapy supervisor, is undertaken before the research commences and continues throughout the duration of the research.
3. The challenge of obtaining free and informed consent in these circumstances is adequately considered ....
4. The impact of the dual relationship is carefully monitored and, when appropriate, addressed in any reports of the research process and outcomes.
5. The use of any records is restricted to the purpose(s) for which they were created and authorised by the client's consent. (Bond 2004, p. 9)

5.2.2   *Beneficence and Nonmaleficence*

As indicated previously, this dual principle states that the benefits of the research are to be maximized and the harms are to be minimized. This principle gives rise to the requirements that the potential risks of the research be outweighed by the potential benefits, that the research design be sound, that the researcher be competent to conduct the proposed research, and that the welfare of the research participants be protected. The hypothetical case below demonstrates how this principle might be unintentionally violated in the context of sandplay therapy.

Case Illustration
A sandplay therapist wishes to evaluate the effectiveness of sandplay therapy. Because the therapist's practice is highly specialized, for example, focused on trauma resulting from sexual abuse, and he wishes to have a more diverse sample of clients, the therapist solicits and collects case studies from various sandplay therapists throughout the country. He is able to garner cases from many sandplay therapists, but does not have a complete listing of all therapists engaged in sandplay therapy for such issues with their clients and does not have a complete database of all clients who have seen the contributing therapists for this issue, because some clients could not be reached or were unwilling to allow their records to be used. The therapist-researcher compiles the cases that are contributed and develops a manuscript that he submits for publication. However, the therapist-researcher does not have a large sample size and is unable to ascertain the extent to which his study sample is representative of the larger population of sandplay clients who have experienced trauma due to sexual abuse. Because of the small number of cases that are included and the method of sampling, the manuscript is ultimately not publishable.

It could be argued that the lack of an adequate sample size and rigorous sampling process has led to an inadvertent breach of the principles of beneficence and nonmaleficence. The participants have not benefited from their participation and there has been no contribution to the general knowledge or understanding. Even with precautions, there is always the possibility that confidentiality may be breached, potentially subjecting the participants to harm. In this situation, the participants were placed at potential risk with no potential benefit. A redesign of the study prior to its initiation to ensure an adequate sample size and more rigorous sampling procedures would avoid this.

5.2.3 Justice

Justice refers to the obligation of the researcher to assist in the fair allocation of resources and burdens. Rawls conceived of differences between individuals in terms of the resources and benefits available to them—"the difference principle"—as

> an agreement to regard the distribution of natural talents as in some respects a common asset and to share in the greater social and economic benefits made possible by the complementarities of the distribution, Those who have been favored by nature, whoever they are, may gain from their good fortune only on terms that improve the situation of those who have lost out. (Rawls 1999, p. 87)

Accordingly, justice has not been effectuated unless:

> All social values—liberty and opportunity, income and wealth, and the social bases of self-respect—are to be distributed equally unless an unequal distribution of any, or all, of these values is to everyone's advantage. (Rawls 1999, p. 54)

The *Belmont Report* (National Commission for the Protection of Human Subjects of Biomedical and Behavioral Research 1979, p. 7–8) noted

> Justice is relevant to the selection of subjects of research at two levels: the social and the individual. Individual justice in the selection of subjects would require that researchers … not offer potentially beneficial research only to some patients who are in their favor.
> Injustice may appear in the selection of subjects, even if individual subjects are selected fairly by investigators and treated fairly in the course of research. This injustice arises from social, racial, sexual, and cultural biases institutionalized in society ….
> Although individual institutions or investigators may not be able to resolve a problem that is pervasive in their social setting, they can consider distributive justice in selecting research subjects.

This raises an important question in the context of sandplay therapy research: Are there differences in the distribution of the benefits and burdens of sandplay research across groups? Many individuals are unable to access sandplay therapists due to a lack of health-care coverage, inadequate health-care coverage, or the unavailability of sandplay therapists in their geographical region. There has been significant discussion in professional circles around the need to outreach to diverse populations who may not have private insurance that covers the cost of sandplay therapy and who may not have the financial means to pay for such services out-of-pocket. (Ethical implications related to access to sandplay therapy in a clinical context are addressed in Chap. 3.) It is often those who are already receiving sandplay therapy who will be invited to participate in sandplay research, via a request to sign a release allowing their case to be presented or reported. Access to sandplay therapy may be a precondition to access to research.

The principle of justice dovetails with the principle of respect for persons in its focus on the development and implementation of special protections for vulnerable research participants, such as those with diminished capacity. For example, the Council of International Organizations for Medical Sciences (2002) has stated,

> Differences in distribution of burdens and benefits are justifiable only if they are based on morally relevant distinctions between persons; one such distinction is vulnerability. "Vulnerability" refers to a substantial incapacity to protect one's own interests owing to such impediments as lack of capability to give informed consent, lack of alternative means of obtaining medical care or other expensive necessities, or being a junior or subordinate member of a hierarchical group. Accordingly, special provision must be made for the protection of the rights and welfare of vulnerable persons.

5.3   Sandplay Research with Children Participants: The Requirement of Assent

Just as in the context of therapy, children are legally unable to provide consent to participate in research and are not presumed to have capacity to consent. However, children's participation in most research generally requires that they provide assent in addition to the informed consent of one or both parents. *Assent* means that the child is aware of the nature of his or her condition, understands what he or she can expect in the context of the research, and indicates his or her willingness to participate in the study. This means that the therapist-researcher must present the child with information about the research in a manner that is developmentally appropriate, assess the extent to which the child understands the information presented, and ascertain whether the child is willing to participate. In most cases, research should not be conducted without the child's assent. In the context of sandplay therapy research, it is difficult to identify a situation in which the benefit of the research would be sufficiently great so as to override a child's unwillingness to participate.

> **Case Illustration**
> A sandplay therapist wishes to report a series of case studies involving the use of sandplay with young teens with a diagnosis of autism. The therapist-researcher approaches a parent of each child for his or her consent; all parents give their consent. The therapist-researcher also requests the assent of the teens. Several of the teens indicate that they do not want their sand trays utilized in the case series and do not want to give assent. Ethically, the therapist-researcher should not include their sand trays in the case series study.

5.4 Other Ethical Considerations

5.4.1 Continuing Consent

Circumstances may change during the course of the study which may impact an individual's willingness or ability to continue with his or her participation and, consequently, require the re-consenting of the individual to assure ongoing validity of his or her consent to participate. As one example, an individual who is experiencing symptoms of early onset dementia may initially be willing to participate in a study relating to the effectiveness of her sandplay therapy. However, as the dementia progresses, she may experience diminished understanding of what exactly is happening. In such a situation, the therapist-researcher may be obliged ethically to obtain informed consent again, in order to ensure that the client has the necessary capacity and understanding to continue as a research participant.

5.4.2 Therapeutic Misconception

Some individuals may also believe that they would not have been offered the possibility of participation in a study unless the researcher believed that their participation would yield some clinical benefit to them personally. They may believe this despite all assertions by the therapist-researcher that they may not receive any personal benefit from their participation and only future patients/clients will derive any benefit from the newfound knowledge gained through the study. This misconception is known as the "therapeutic misconception" (Grisso and Appelbaum 1998).

5.4.3 After the Research: Publication and Dissemination

The therapist-researcher's ethical responsibilities continue even after the conclusion of the research. Mental health professional associations across various countries indicate that ethical research requires that the therapist-researcher report the research findings accurately and continue to preserve the confidentiality and privacy of the research participants (American Psychological Association 2014; Australian Psychological Society 2007; Bond 2004; National Association of Social Workers 2008). A variety of strategies can be utilized to protect the identity of the research participants and safeguard the confidentiality of their individual data. These include aggregating the data from multiple individuals, excluding identifying descriptions of individuals, and conflating multiple accounts or scenarios into one representative account or case study (Bond 2004; National Association of Social Workers 2008).

It is also important that the therapist-researcher acknowledge the contributions of others to his or her research. In general, individuals who have made a significant intellectual contribution to the research should be acknowledged as coauthors.

Others who assisted with the research, such as an assistant for the transcription of recorded interviews or for data entry, can be thanked in an acknowledgment section (Wager and Kleinert 2010; International Committee of Medical Journal Editors 2014; National Association of Social Workers 2008).

5.5   Conclusion

There is an increasing need and interest in the conduct of sandplay-related research in order to evaluate the effectiveness and efficacy of this modality. Concurrent with the conduct of the research, greater attention must be paid to the associated ethical issues. Although the provisions of the Nuremberg Code, the Helsinki Declaration (World Medical Association, 2013), and the Council of International Organizations for Medical Sciences were developed for the conduct of biomedical research, their provisions are relevant to the conduct of psychological research. Similar provisions exist across numerous countries to guide mental health professionals engaging in psychological research, such as those promulgated by the APA, the Australian Psychological Society, the British Association for Counselling and Psychotherapy, the British Psychological Society, the National Association of Social Workers (the USA), and the Psychological Society of Ireland.

Ethical research in sandplay requires that the therapist-researcher ensure that the research procedures, including the informed consent and enrollment processes, reflect the ethical principles of respect for persons, beneficence nonmaleficence, and justice. In so doing, the investigator must provide the prospective participant with information sufficient and adequate to permit him or her to decide whether to participate and minimize any associated risks to the prospective participant both during and after the conclusion of the research. The therapist-researcher is also challenged to assess whether there exist any conflicts of interests in pursuing the research and whether he or she is competent to conduct the contemplated research. Situations in which the therapist-researcher is providing therapy in addition to conducting the research may demand more intensive supervision of the researching therapist.

References

Achenbach, T. M. (1991). *Manual for the teacher's report form and 1991 profile*. Burlington: Department of Psychiatry, University of Vermont.

American Psychological Association. (2014). *Ethical principles of psychologists and code of conduct including 2010 amendments*. Washington, D.C.: American Psychological Association. http://www.apa.org/ethics/code/index.aspx?item!!/span>=11. Accessed23 June 2014.

Ammann, R. (1991). *Healing and transformation in sandplay*. LaSalle: Open Court.

Auriti, E. (2013). Who can obtain access to research data? Protecting research data against compelled disclosure. *NACUA Notes, 11*(7). Washington, D.C.: National Association of College and University Attorneys. https://www.calstate.edu/gc/documents/ NACUANOTES-Who-CanObtainAccess-to-Research-ProtectingData.pdf. Accessed 19 June 2014.

Australian Psychological Society. (2007). *APS code of ethics*. Melbourne: Australian Psychological Society Limited. http://www.psychology.org.au/Assets/Files/APS-Code-of-Ethics.pdf. Accessed 23 June 2014.

Bond, T. (2004). *Ethical guidelines for researching, counseling and psychotherapy*. London: British Association for Counselling and Psychotherapy.

British Psychological Society. (2010). *Code of human research ethics*. Leicester: British Psychological Society.

Canadian Psychological Association. (2010). *Canadian code of ethics for psychologists, third edition*. Ottawa: Canadian Psychological Association. http://www.cpa.ca/cpasite/userfiles/Documents/Canadian%20Code%20of%20Ethics%20for%20Psycho.pdf. Accessed 23 June 2014.

Carter, A., & Hall, W. (2008). The issue of consent in research that administers drugs of addiction to addicted persons. *Accountability in Research, 15*, 209–225.

Charland, L. C. (2002). Cynthia's dilemma: Consenting to heroin prescription. *American Journal of Bioethics, 2*, 37–47.

Council of International Organizations for Medical Sciences. (2002). *International ethical guidelines for biomedical research involving human subjects*. Geneva: Council of International Organizations for Medical Sciences.

Council of International Organizations for Medical Sciences. (2009). *International ethical guidelines on epidemiological studies*. Geneva: Council of International Organizations for Medical Sciences.

Davis v. Lhim. (1983). 124 Mich. App. 291, *aff'd on rem* 147 Mich. App. 8 (1985), *rev'd* on grounds of government immunity in *Canon v. Thumudo*, 430 Mich. 326 (1988).

De Craemer, W. (1983). A cross-cultural perspective on personhood. *Milbank Memorial Fund Quarterly, 61*, 19–34.

De Vries, R., DeBruin, D. A., & Goodgame, A. (2004). Ethics review of social, behavioral, and economic research: Where should we go from here? *Ethics & Behavior, 14*(4), 351–368.

Dresser, R. (2001). Advance directives in dementia research. IRB: *Ethics and Human Research, 23*(1), 1–6.

Earle v. Kuklo. (1953). 26 N.J. Super. 471 (App. Div.).

Etherington, K. (2007). Ethical research on reflexive relationships. *Qualitative Inquiry, 13*(5), 599–616.

Exner, J. E. (1986). *The Rorschach: A comprehensive system : Basic foundations* (Vol. 1) New York: Wiley.

Faden, R. R., & Beauchamp, T. L. (1986). *A history of informed consent*. Oxford: Oxford University Press.

Grisso, T., & Appelbaum, P. (1998). *Assessing competence to consent to treatment: A guide for physicians and other health professionals*. New York: Oxford University Press.

Hayes, C. 2011 (September 2). IRA researchers at Boston College file suit against US govt. Irish Central. http://www.irishcentral.com/news/others-from-boston-college-project-file-separate-suit-to-suppress-ira-interviews-129168208–237409721.html. Accessed 19 June 2014.

Hong, G. L. (2011). *Sandplay therapy: Research and practice*. New York: Routledge.

International Committee of Medical Journal Editors. (2014). Defining the role of authors and contributors. http://www.icmje.org/recommendations/browse/ roles-and-responsibilities/defining-the-role-of-authors-and-contributors.html. Accessed 19 June 2014.

Jablonski v. United States. (1983). 712 F.2d 391 (9th Cir.).

Kalff, D. M. (1980). *Sandplay: A psychotherapeutic approach to the psyche*. Boston: Sigo.

Kovacs, M. (1979). *Children's Depression Inventory*. Pittsburgh, PA: University of Pittsburgh.

Levine, R. J. (1988). *Ethics and regulation of clinical research*. New Haven: Yale University Press.

Ling, W. (2002). Cynthia's dilemma. *American Journal of Bioethics, 2*, 55–56.

Lundqvist, T. (2005). Cognitive consequences of cannabis use: Comparison with abuse of stimulants and heroin with regard to attention, memory and executive functions, *Pharmacological and Biochemical Behavior, 81*, 319–330.

McIntosh v. Milano. (1979). 168 N.J. Super. 466.

National Association of Social Workers. (2008). Code of ethics. Available at: http://www.social-workers.org/pubs/code/code.asp. Accessed 7 July 2014.

National Bioethics Advisory Commission. (1998). *Research involving persons with mental disorders that may affect decisionmaking capacity. Vol. I: Report and recommendations of the National Bioethics Advisory Commission.* Rockville, MD: U.S. Government Printing Office.

National Bioethics Advisory Commission. (2001). *Ethical and policy issues involving human participants* (Vol. 1). Rockville: U. S. Government Printing Office.

National Commission for the Protection of Human Subjects of Biomedical and Behavioral Research. (1979). The Belmont Report: Ethical principles and guidelines for the protection of human subjects of research. Washington, D.C.: United States Department of Health, Education, and Welfare [DHEW Pub. No. OS 78–0012].

Nuremberg Code. (1946). In K. Lebacqz & R. J. Levine. (1982). Informed consent in human research: Ethical and legal aspects. In W. T. Reich (Ed.), *Encyclopedia of bioethics* (p. 757). New York: Free press.

People v. Newman. (1973). 298 N.E.2d 651 (App. Div.).

Perring, C. (2002). Resisting the temptations of addiction rhetoric. *American Journal of Bioethics, 2,* 51–52.

Psychological Society of Ireland. (2011). *Code of professional ethics.* Dublin: Psychological Society of Ireland.

Rawls, J. (1999). *A theory of justice* (rev. ed.) Cambridge: The Belknap.

Tarasoff v. Regents of the University of California. (1976). 17 Cal. 3d 425.

Thompson v. County of Alameda. (1980). 27 Cal. 3d 741.

Wager, E., & Kleinert, S. (2010). Responsible research publication: International standards for authors. A position statement developed at the Second World Conference in Research Integrity, Singapore, July 22–24. http://publicationethics.org/files/International%20standards_authors_for%20website_11_Nov_2011.pdf. Accessed 19 June 2014.

Weinrib, E. L. (1983). *Images of the self: The sandplay therapy process.* Cloverdale: Temenos.

World Medical Association. (2013). Declaration of Helsinki: Ethical principles for medical research involving human subjects, 64th WMA General Assembly. Fortaleza, Brazil.

Chapter 6
Dual Relationships and Conflict of Interest in Sandplay Therapy

Sana Loue and Jean Parkinson

6.1 Dual Relationships Between Therapist and Client

Consider the following scenario.

> **Scenario 1**
>
> A sandplay therapist provides services gratis to minority individuals who lack health insurance coverage. Many of these individuals would otherwise be unable to obtain services, due to both the lack of insurance coverage and other circumstances in their lives, such as lack of available transportation, unavailability of any mental health services at community agencies outside of the usual workday hours, and the relative absence of therapists in that geographic area who utilize sandplay therapy. The therapist is active in the community and, as a result, often encounters clients at grocery stores, social functions, and community forums. These encounters with clients are not planned.

Termination of the therapeutic services with these clients would likely result in the clients' inability to access sandplay therapy, raising the ethical issue of distributive justice. The therapist's withdrawal from the community activities could endanger the trust that he/she has developed with the community. Yet, the existence of an additional relationship between the therapist and the client—a dual relationship—raises ethical issues that must be addressed:

S. Loue (✉)
Case Western Reserve University, Cleveland, OH, USA
e-mail: sana.loue@case.edu

J. Parkinson
Auckland, New Zealand

© Springer International Publishing Switzerland 2015
S. Loue (ed.), *Ethical Issues in Sandplay Therapy Practice and Research,*
SpringerBriefs in Social Work, DOI 10.1007/978-3-319-14118-3_6

A dual relationship between a therapist and the client occurs when: [a] professional assumes
a second role with a client, becoming social worker and friend, employer, teacher, business
associate, family member, or sex partner. A practitioner can engage in a dual relationship
whether the second relationship begins before, during, or after the [professional therapeu-
tic] relationship. (Kagle and Giebelhausen 1994, p. 213)

Dual relationships appear to be one of the most vexing areas of practice for mental
health care providers. A survey of 679 US psychologists that focused on challenging
or troubling ethical issues that they encountered in practice revealed that 17 % of
such instances related to blurred, dual, or conflictual relationships (Pope and Vetter
1992). An additional 4 % of the 703 reported instances concerned unethical sexual
involvements. Dual relationships constitute a large proportion of complaints raised
in malpractice claims, disciplinary actions, and ethics complaints (Ethics Committee of the American Psychological Association 1988; Pope 1989a, b).

A dual relationship may potentially be harmful or beneficial to a client, or may
have no discernible impact on the client or the course of therapy. Dual relationships
are—or should be—of concern to therapists precisely because of the risk of harm to
the client and/or the therapeutic relationship. The ethical principle of beneficence
requires that the health care professional strive to maximize the benefit to the client, while the converse principle, nonmaleficence, dictates that the professional act
to avoid potential harm to the client. As Moleski and Kiselica observed in their
discussion of moral principles relevant to dual relationships, "In a dual relationship, the degree of potential for destructiveness is relative to the potential degree
of autonomy lost by the client" (Moleski and Kiselica 2005, p. 4). The harm may
occur because the dual relationship requires that both the therapist and the client
assume different roles than they would have in the context of therapy. This change
of roles may lead to confusion and a loss of objectivity on the part of the client and/
or the therapist (Moleski and Kiselica 2005; Pipes 1997). A discussion of such role
changes in the context of a therapeutic session may become necessary to identify
and address any resulting confusion (Sterling 1992).

Dual relationships are of particular concern in the context of sandplay practice
due to the relative lack of certified sandplay therapists. Many clients would prefer
to seek therapy from the most highly trained sandplay therapists. Outside of major
urban areas such as New York and San Francisco, it is likely that, given the relative
scarcity of certified sandplay therapists and the smaller size of the communities in
which many sandplay therapists practice, a relationship may already exist in some
form between the therapist and the prospective client. This relationship may be
relatively superficial, such as sitting in adjoining seats at the symphony, or may
involve greater depth, such as serving on a committee together at their children's
elementary school.

That said, under some circumstances a refusal to provide services due to a
dual relationship may also raise ethical concerns (cf. Doyle 1997). Such a refusal
might contravene the ethical principle of justice, as well as those of beneficence

and nonmaleficence. (For a more detailed discussion of the principle of justice, see Chap. 3, which addresses access to sandplay therapy.) Such a situation might arise, for example, if a sandplay therapist whose practice is the only sandplay practice in the geographic area and whose practice is open to new clients refuses to accept a child as a client because his or her own child attends the same school. Maintaining a boundary to prevent a dual relationship in such a situation may result in an unnecessary emphasis on the power differential inherent in the therapeutic relationship (Moleski and Kiselica 2005).

Clearly, the decision to engage in a dual relationship with a client must be weighed carefully. The *Code of Ethics* of the International Society of Sandplay Therapists duly cautions against the therapist's nonprofessional relationships with clients, the clients' romantic partners, and family members (International Society of Sandplay Therapists 2007, par. Chap. 1). The therapist must consider and weigh the potential harm to the client that may result from the dual relationship against the potential harm that may flow from the rejection of the dual relationship (Corey et al. 1998).

6.2   Dual Relationships Between Sandplay Therapists

Issues relating to dual relationships also confront therapists aspiring to become certified in the practice of sandplay therapy. In the USA, for example, certification in sandplay therapy requires a minimum of a 40-h personal process and 80 h of supervision/consultation, both of which must be pursued with a certified sandplay therapist (Sandplay Therapists of America 2012). Additionally, the applicant for certification must have completed two preliminary papers, each reviewed and approved by a certified teaching sandplay therapist and final case study, reviewed and approved by three different certified teaching sandplay therapists. Accordingly, the process of sandplay certification requires the review and approval at various stages of a minimum of seven certified therapists. An even greater number may be required in situations in which there are no certified sandplay therapists in the applicant's geographic area and he or she must utilize opportunities for consultation while attending conferences at distant locations. Because the process of certification may take years—in some cases, a decade—and require the participation of a relatively large proportion of individuals from a relatively small pool, it is likely that a dual relationship would exist at some level between the applicant and one or more reviewers by the time the certification process has been completed. This may occur as the result of professional interactions at conferences and unexpected meetings at events sponsored by mutually known colleagues.

The following scenario also illustrates how a dual relationship might occur between sandplay therapists and why the role changes required by such dual relationships may be problematic.

Scenario 2

A sandplay therapist resides in a geographic area in which there are few sand-play therapists and none who are certified in sandplay therapy. She wishes to undergo her own process, but also requires consultation for the cases that she is currently following. As a consequence, she must travel a great distance at considerable expense to undergo her personal process and to obtain consultation. She agrees with a second sandplay therapist who is certified that they will exchange consultation services with each other, in other words, barter consultation services, and that she will also obtain counseling services from this second therapist.

This type of arrangement, by which the two therapists (1) are engaged in a dual relationship as therapists and as client–therapist and (2) are bartering consultation services is explicitly prohibited by many ethical codes governing mental health practice. The American Psychological Association provides:

> Barter is the acceptance of goods, services, or other nonmonetary remuneration from clients/patients in return for psychological services. Psychologists may barter only if (1) it is not clinically contraindicated, and (2) the resulting arrangement is not exploitative. (American Psychological Association 2010, p. 9, par. 6.05)

In this situation, it could reasonably be argued that the bartering relationship between the therapists may be clinically contraindicated because one therapist is also seeking therapy from the other. Difficulties or tensions within the bartering relationship and, indeed, marked differences in how the therapists work with and evaluate their cases, may color the personal process relationship.

6.3   Characterizing Dual Relationships

Boundaries are necessary to

> mark the limits or parameters of appropriate, good, and ethical practice, including both structural (e.g., roles, time, place-space) and process (e.g., gifts, language, self-disclosure, physical contact, interactional patterns) dimensions. (Lamb and Catanzaro 1998, p. 498)

Some dual relationships may be thought of as boundary violations; these often involve exploitation or coercion of the client. Examples include sexual relations with a client and influencing a terminally ill client to bequeath a large portion of his or her assets to the mental health care provider (Reamer 2003). Such dual relationships are almost invariably harmful to the client. A dual relationship involving sexual contact may lead the client to experience cognitive dysfunction, guilt, anxiety, depression, sexual confusion, and suicidal ideation (Kagle and Giebelhausen 1994; Smith and Fitzpatrick 1995; Stake and Oliver 1991). The client may come to distrust all health care professionals (Stake and Oliver 1991).

Additionally, dual relationships that violate boundaries often constitute a conflict of interest (discussed further below), and are generally believed to be unethical. As an example, the *Ethical Principles of Psychologists and Code of Conduct* of the American Psychological Association states: "Psychologists do not exploit persons over whom they have supervisory, evaluative, or other authority such as clients/patients, students, supervisees, research participants, and employees" (American Psychological Association 2010, p. 6, par. 3.08). The document specifically prohibits sexual relationships between the therapist and a current client, between a therapist and "close relatives, guardians, or significant others of current clients/patients. Psychologists do not terminate therapy to circumvent this standard" (American Psychological Association 2010, p. 14, par. 10.06). Psychologists are also prohibited from engaging in sexual relations with past clients if less than 2 years have transpired since the termination of therapy and from accepting as clients any individuals with whom they have previously had sexual intimacies. A similar, and even more rigorous standard, governs the conduct of many US-based clinical social workers: "Clinical social workers do not, under any circumstances, engage in either romantic or sexual conduct with either current or former clients" (Clinical Social Work Association n.d., par. 3).

In contrast to those dual relationships that constitute boundary violations, boundary crossings are not intentionally coercive or exploitative and may, at times, be helpful. As an example, a therapist may reveal details about his or her life to a client in an effort to encourage the client in his or her therapeutic process. The revelation may, indeed, help the client or it may confuse the client due to transference issues, although the therapist's action was guided by an intent to be helpful to the client. (It is acknowledged here that therapists that utilize and are faithful to the Kalffian approach to sandplay are not likely to engage in such self-revelations, but the possibility of such an occurrence nevertheless exists.) A therapist may agree to attend a client's high school graduation as a way to demonstrate his or her caring and the importance of the client's achievement.

A dual relationship would be established and a boundary violation would exist if, as in the second scenario above, a therapist seeking certification in sandplay therapy were to undergo both his or her personal process and obtain his or her consultation hours with the same sandplay therapist fulfilling both the roles of therapist and consultant. Whether the dual relationship would be harmful to the aspiring sandplay therapist would depend on any number of factors specific to the situation. Indeed, any harm or benefit from such an arrangement might not be known until adequate time had passed to enable the aspirant to look back on his or her experience. The Sandplay Therapists of America recognizes the risks inherent in such dual relationships involving therapists aspiring to certification and, for this reason, does not permit such a dual relationship. However, dual relationships may exist if only because of unanticipated circumstances.

Dual relationships exist along a spectrum of both intensity and intentionality. At one end of the spectrum, a sandplay therapist might encounter a client by chance shopping at the same supermarket on a single occasion. In such an instance, the intensity of the dual relationship is extremely low and the encounter occurred in the absence of any planning by either the therapist or the client. At the other end of the spectrum, illustrating both great intensity and intentionality, is the case of a therapist who becomes romantically involved with the client's school guidance counselor who referred the child to the therapist and who continues to see the child in his school.

Many professional codes of ethics recognize the spectrum along which dual relationships may occur and the need to avoid them where possible. As an example, the *Code of Professional Ethics* of the Psychological Society of Ireland provides that a psychologist should

> Be acutely aware of the problematic nature of dual relationships (with, for example, students, employees or clients), and recognise that it is not always possible to avoid them (for example, when offering services in a small community, or engaging in person-centred teaching or training). Where it is possible, psychologists shall avoid such relationships; where it is not, they take active steps to safeguard the students', employees' or clients' interests. (Psychological Society of Ireland 2010, p. 15, par. 4.4.1)

Reamer (2003) has suggested that dual relationships can be classified into five types based on the nature or purpose of the relationship. *Intimate relationships* encompass sexual relations and physical contact. Intimate relationships that involve sexual relations between a therapist and current or former client are generally prohibited by ethical codes governing the conduct of mental health care providers and have been discussed extensively in the extant literature. Accordingly, that discussion is not repeated here. Rather, the interested reader is referred to various other sources for this discussion (Pope and Vetter 1992; Lamb and Catanzaro 1998). At least one author has suggested that any physical contact between the therapist and the client is classifiable as a dual relationship even while acknowledging that some physical contact may be nonsexual in nature and, in many cases, is unlikely to be harmful to the client (Reamer 2003).

Relationships entered into for the *benefit of the therapist* may assume the form of bartering, such as exchanging counseling services for car repairs or seeking information from the client, such as asking a client for stock tips. Some dual relationships may be designed to *meet the emotional and dependency needs of the therapist,* e.g., reliance on the client for companionship at the mall. Dual relationships may be premised on *altruism,* such as the giving and receiving of gifts between the therapist and the client.

Finally, interactions that constitute dual relationships may result from *unanticipated circumstances* that are often beyond the control of the therapist, such as attendance at the same place of worship. These situations are more likely to occur in smaller communities or in situations in which the therapist plays an active role in the community in order to establish his or her credibility, engender trust, and gain acceptance (Helbok 2003). The *Canadian Code of Ethics for Psychologists* advises that psychologists who find themselves in such situations should

> [m]anage dual or multiple relationships that are unavoidable due to cultural norms or other circumstances in such a manner that bias, lack of objectivity, and risk of exploitation are minimized. This might include obtaining ongoing supervision or consultation for the duration of the dual or multiple relationship, or involving a third party in obtaining consent

(e.g., approaching a client or employee about becoming a research participant). (Canadian Psychological Association 2000, p. 27, par. III.34)

Considerable controversy surrounds the issue of dual relationships after the therapeutic relationship has terminated. As noted above, some codes of ethics prohibit all sexual relations between therapists and their former clients, regardless of the length of time that has passed since the termination of the therapeutic relationship. Other codes permit such relations after a specified period of time has transpired.

Similar disagreement exists with respect to the acceptability and advisability ethically of nonsexual dual relationships after therapy has terminated (Salisbury and Kinnier 1996; Reamer 2003). Like dual relationships that occur during the course of the therapeutic relationship, post-therapy dual relationships may be intentional or circumstantial (Anderson and Kitchener 1996). Most frequently, such relationships take the form of friendships or personal relationships and, somewhat less frequently, as a business/financial, collegial/professional, or supervisory/evaluative relationship. One author observed that

> some social workers argue that friendships with former clients are not inherently unethical and reflect a more egalitarian, nonhierarchical approach to practice. These social workers typically claim that emotionally mature social workers and former clients are quite capable of entering into new kinds of relationships following termination of the professional-client relationship and that such new relationships often are, in fact, evidence of the former client's substantial therapeutic progress. (Reamer 2003, p. 126)

In the context of sandplay therapy, transformation of the therapist–client relationship would not be uncommon. For example, a sandplay therapist seeking certification and undergoing his or her personal process might later form a mutually beneficial peer relationship with the therapist that he or she saw for the personal process. In yet another scenario, the sandplay therapist engaged for the personal process might later join an agency where he or she is supervised by the former client-therapist. It would be critical under any of these circumstances to explore together the basis for the new relationship. There exists the possibility that knowledge acquired during the course of the therapeutic relationship could affect each individual's ability to remain objective in their new roles (Anderson and Kitchener 1996).

A thoughtful, ethical therapist will engage the client in a discussion to address the potential implications of a dual relationship before embarking on such a course. A discussion of the boundaries of the therapeutic process is paramount and should be considered within the context of the informed consent process (Corey et al. 1998). That process should also address the potential risks and benefits that may accompany the contemplated dual relationship. During the course of the therapeutic relationship, the therapist and the client should be prepared to discuss any issues or challenges that arise because of the dual relationship. The therapist would be wise to seek consultation from a more experienced practitioner regarding the dual relationship and its impact on the therapeutic relationship and be willing to refer the client to another professional should that be deemed necessary or advisable. The dual relationship should be noted in the case notes. Finally, it is critical that the therapist examine his or her own motives that may underlie the decision to participate in a dual relationship (Corey et al. 1998).

It has been suggested that the following factors be considered when evaluating the advisability of entering a nontherapeutic relationship with a client following the termination of therapy: The length of time that has elapsed since therapy was terminated, transference issues, the length and nature of the therapy, the nature of the termination, freedom of choice, whether exploitation occurred during the therapeutic relationship, the client's current mental health status, the possibility that therapy will be reactivated with the client, and the potential for harm to the client (Akamatsu 1988).

We suggest, as well, that the therapist engage in a self-evaluative process to identify issues of countertransference and the basis of his or her desire to engage further with the former client. Prior to assuming new roles outside of those of the therapist and client, it will be important to discuss openly with the client the probability that the client will not be able to reengage in therapy with that therapist if they assume the new, contemplated extra-therapy roles.

6.4   Conflicts of Interest

A conflict of interest has been defined as a situation

> that can lead to distorted judgment and can motivate psychologists to act in ways that meet their own personal, political, financial, or business interests at the expense of the best interests of members of the public. (Canadian Psychological Association 2000, p. 23)

Case Example 1

A psychologist has been providing therapy to a middle-aged woman who has been diagnosed with bipolar disorder and alcoholism. The client is only intermittently adherent to her prescribed medication regimen. During the course of therapy, she has been hospitalized on several occasions following suicide attempts. On one occasion, she was involved as the driver in a hit-and-run accident with a pedestrian. Both the psychologist and the client are unhappy in their marriages. The psychologist frequently shares stories about her marital difficulties with the client. After several years of therapy with the therapist, during which time much such sharing has occurred, the therapist advises the client that she does not believe the client is an alcoholic and does likely not have bipolar disorder, diagnoses that the client has never accepted. The client's husband is well-off financially. The psychologist proposes to the client that they develop a gift card business together and that the client's husband provide the start-up cash.

Although many conflicts of interest arise from dual relationships, some do not. As an example of a conflict of interest that is not a dual relationship, consider a situation in which a client is concerned about the impending sale of a company that may lead to his friend's employment termination. Unbeknownst to the client, the therapist owns stock in the company to be sold. Use by the therapist of this insider knowledge to buy or sell his stock to maximize his profit would violate client confidentiality, exemplifying a conflict of interest between the obligation owed to the client and the therapist's own financial interest.

According to the Canadian Code of Ethics for Psychologists,
It is the responsibility of psychologists to avoid dual or multiple relationships and other conflicts of interest when appropriate and possible. When such situations cannot be avoided or are inappropriate to avoid, psychologists have a responsibility to declare that they have a conflict of interest, to seek advice, and to establish safeguards to ensure that the best interests of members of the public are protected. (Canadian Psychological Association 2000, p. 23)

Similarly, the Australian and New Zealand Arts Therapy Association (ANZATA) provides in its *Standards of Professional Practice and Code of Ethics:*

Arts Therapists are responsible for setting and maintaining appropriate professional boundaries. This includes avoiding any situations that compromise a sense of objectivity, and/or presents a conflict of interests. They must not engage in dual relationships (e.g. personal or business relationships with clients). (Australian and New Zealand Arts Therapy Association n.d., par. 9)

Codes of ethics for many mental health professionals—psychologists, social workers, counselors—in many countries contain similar proscriptions and advisories (American Psychological Association 2010; Australian Psychological Society n.d.; Clinical Social Work Association n.d.; Psychological Society of Ireland 2010; Social Workers registration Board 2014).

Just as dual relationships may occur between a therapist and client and between therapists themselves, so too may conflicts of interest arise between a therapist and his or her client or between two therapists. A dilemma arises as to the adequacy of supervision if the supervisor is not a registered member of the same profession or does not have experience with the primary client group(s) with whom the supervisee is working. A further conflict of interests can arise when the supervisor has responsibility for institutional/line management/appraisal issues in relation to the supervisee. This systemic priority can result in inadequate time for clinical work. (For a detailed discussion of such issues, see Chap. 1 of this text.)

Case Example 2

A therapist aspiring to certification in sandplay therapy (therapist-client) is undergoing her personal process with a certified sandplay therapist (certified therapist). The certified therapist is charging the therapist-client US$ 125 per hour for her personal process. Certification requires that individuals undergo a minimum of 40 h of personal process. There is an inherent conflict of interest in the relationship because the certified therapist stands to benefit financially from the therapist-client's continued personal process. The certified therapist has the authority and power to decide that 40 h is inadequate and the therapist-client is not ready to proceed further in the process absent additional therapy.

6.5   Consultation and Supervision

Ethical issues specific to cyber-supervision in sandplay are discussed in detail in Chap. 2 of this text. Here, we focus on the distinctions between consultation and supervision, the differing ethical and sometimes legal issues associated with each, and ethical challenges associated with consultation in the context of sandplay training and practice.

6.5.1   The Consultant Relationship

Consultation with colleagues can take any one or more of several forms: evaluation of another therapist's client, the provision of assistance to a therapist for the purpose of developing a client's treatment plan or providing general support with respect to the client, or the provision of services at an administrative or organizational level to a private practice or an agency (Clayton and Bongar 1994). Consultation may occur formally as reflected in a written agreement and fee for the consultation service, or on a more informal basis. It may occur between members of the same mental health profession, e.g., psychologists, or may be interprofessional, such as between a psychologist and a psychiatrist (Sweet and Rozensky 1991). Consultation with other, more experienced clinicians may minimize therapist stress, provide additional insights into the dynamics of a therapeutic relationship, diminish therapist cognitive and affective bias, and reduce the likelihood that clients will be dissatisfied with their therapeutic experience (Gutheil 1990; Pope et al. 1987; Risley and Sheldon-Wildgen 1982). Consultation may be a particularly critical component of risk management and prevention when providing services to clients who are suicidal (Bongar 1991) or who are engaging in self-injurious behaviors that may potentially lead to inadvertent suicide, e.g., severe anorexia, self-mutilating behaviors. Indeed, ethical codes governing mental health professionals in various countries explicitly recommend that practitioners seek out consultation services in specified situations, e.g., to more effectively provide services, when they are experiencing personal problems that may interfere with their work-related responsibilities, and when they are seeking to provide new services or services to a population that is new to them (American Psychological Association 2010; Australian Psychological Society n.d.; Social Workers Registration Board 2014).

Ethical issues—and sometimes legal issues—may arise between the therapist seeking consultation and the therapist providing those consultation services. These issues may be particularly challenging in the context of sandplay therapy consultation because the therapists seeking and providing the consultation may belong to different mental health professions, each with its own code of ethics, and may be practicing in completely different legal jurisdictions, e.g., state or country, and so may be subject to differing legal standards. Consider the following composite case example.

Case Example 3

A more senior sandplay therapist who is licensed as a psychologist in one state provides face-to-face consultation services to a more junior sandplay therapist who is licensed as a marriage and family counselor in a second state. The consultation services are provided in a third state, while they are both attending a conference. The consultant-therapist demands detailed information about the client in writing. In order to better understand and address potential cotransference issues between him- or herself and the client, the more junior sandplay therapist discloses highly personal information about him- or herself. The consultant-therapist has assured the individual that all information communicated in the context of consultation is confidential, although there is no written agreement to that effect. The consultant-therapist later discloses to other sandplay therapists portions of what the more junior individual has confided in the context of consultation, disclosing that therapist's identity in the process. When questioned about his/her action, the consultant-therapist justifies the disclosure by indicating that she does not believe that the more junior individual's practice is adequate.

This case example raises the ethical issue of a breach of confidentiality, both between the sandplay therapist providing direct client services through the disclosure of detailed information to the consultant, and between the consultant-therapist and the treating sandplay therapist through the disclosure of information that was understood by the junior therapist to be confidential. Whether either can be considered a violation of a professional code of ethics depends upon the code under which each is governed. Further, depending upon the details underlying the consultant-therapist's conclusion that the junior therapist's treatment provided to the client is inadequate, she may be under an ethical obligation pursuant to her respective professional code to report that individual to the appropriate licensing board. These ethical issues are rendered even more complex because some legal jurisdictions may require the reporting of a colleague's deficient practice to the relevant governing authority, so that a failure to do so may be not only an ethical issue, but a legal one as well.

6.5.2 Consultation and Obligations to the Client

In a consulting relationship, it is the sandplay therapist providing the direct services to the client who is legally responsible; he or she can accept or reject the recommendations of the consultant. This means that the sandplay therapist must continue to adhere to the ethical principles governing the therapist–client relationship, notwithstanding the fact that some information must be shared with the consultant as part of the consultative process. As an example, the American Psychological Association advises:

> When consulting with colleagues, (1) psychologists do not disclose confidential information that reasonably could lead to the identification of a client/patient, research participant, or other person or organization with whom they have a confidential relationship unless they have obtained the prior consent of the person or organization or the disclosure cannot be avoided, and (2) they disclose information only to the extent necessary to achieve the purposes of the consultation. (American Psychological Association 2010, p. 7, par. 4.06)

Accordingly, the psychologist must weigh carefully how much and what information should be disclosed in order to obtain reliable consultative advice while still maintaining the client privacy. Various writers and ethical codes for a number of mental health professions highly recommend and emphasize the importance of discussing the possible need for consultative services at the outset with each client as part of the informed consent process and documenting the client's agreement (American Psychological Association 2010; Clayton and Bongar 1994; Social Workers Registration Board 2014).

Traditionally, the term supervision has been utilized to refer to the relationship between a licensed supervisor and a prelicense trainee working under the direction of the supervisor. It has been asserted that the term has evolved to refer to "a cooperative activity between supervisor and supervising therapist" and now encompasses both supervision and consultation (Friedman and Mitchell 2008, p. 8). That said, the ethical and legal responsibilities of a consultant vis-à-vis the client are quite different from those of a supervisor. A supervisor may be held to a higher standard of care with respect to the particular client than a consultant would likely be. In fact, the supervisor may be liable for the decisions and actions of the treating sandplay therapist under the legal doctrine of *respondeat superior,* which holds supervisors and employers responsible for the actions of their supervisees and employees under specified conditions (Recupero and Rainey 2007). This is true even if the supervisor at an agency has been hired to supervise in the capacity of a consultant rather than an agency employee (Harrar et al. 1990). Consultants may also become liable for the actions and decisions of the treating sandplay therapist if the consultant forms a direct relationship with the client and the client believes that a therapist–client relationship has been established with the consultant (Stromberg et al. 1988).

Clearly, consultation with colleagues may be critical to maintaining one's skills and one's objectivity during the course of providing sandplay therapy. In order to avoid potential ethical and legal challenges, we highly recommend that sandplay practitioners seeking consultation obtain written informed consent from their clients. That consent should provide the client with a general explanation of what consultation is, its purpose, and the risks and benefits that may be associated with consultation.

It is also important that the sandplay therapist seeking consultation have a written agreement with the individual providing such consultation. That agreement should explain in detail various elements of the agreement, including (1) the frequency and duration of the agreement, (2) the scope of services to be provided, (3) the cost of the consultation service, (4) the extent to which confidentiality and privacy of the individual seeking consultation will be maintained, (5) the extent to which details about the client to be discussed will be released to the consultant and the extent to which confidentiality and privacy of the client(s) under discussion will be maintained, and (6) the length of time, format of, and location in which the consultant will maintain records pertaining to the client and to the therapist seeking consultation services.

References

Akamatsu, T. J. (1988). Intimate relationships with former clients: National survey of attitudes and behavior among practitioners. *Professional Psychology: Research and Practice, 19,* 454–458.

American Psychological Association. (2010). Ethical principles of psychologists and code of conduct. Washington, D.C.: American Psychological Association. http://www.apa.org/ethics/code/principles.pdf. Accessed 7 Sept 2014.

Anderson, S. K., & Kitchener, K. S. (1996). Nonromantic, nonsexual posttherapy relationships between psychologists and former clients: An exploratory study of critical incidents. *Professional Psychology: Research and Practice, 27,* 59–66.

Australian and New Zealand Arts Therapist Association. (n.d.) Standards of professional practice and code of ethics. Glebe, Australia: Australian and New Zealand Arts Therapist Association. http://www.anzata.org/assets/Uploads/professional/profinfo/ethicsstandards.pdf. Accessed27 Sept 2014.

Australian Psychological Society. (n.d.). APS code of ethics. Melbourne: Australian Psychological Society Limited. http://www.psychology.org.au/Assets/Files/APS-Code-of-Ethics.pdf. Accessed 7 Sept 2014.

Bongar, B.M. (1991). *The suicidal patient: Clinical and legal standards of care.* Washington, D.C.: American Psychological Association.

Canadian Psychological Association. (2000). Canadian code of ethics for psychologists (3rd edn.). Ottawa: http://www.cpa.ca/docs/File/Ethics/cpa_code_2000_eng_jp_jan2014.pdf. Accessed 7 Sept 2014.

Clayton, S., & Bongar B. (1994). The use of consultation in psychological practice: Ethical, legal, and clinical considerations. *Ethics & Behavior, 4*(1), 43–57.

Clinical Social Work Association. (n.d.). Code of ethics. http://associationsites.com/CSWA/collection/Ethcs%20Code%20Locked%2006.pdf. Accessed 7 Sept 2014.

Corey, G., Corey, M. S., & Callanan, P. (1998). *Issues and ethics in the helping professions.* Pacific Grove: Brooks/Cole.

Doyle, K. (1997). Substance abuse counselors in recovery: Implications for the ethical issue of dual relationships. *Journal of Counseling & Development, 75,* 428–432.

Ethics Committee of the American Psychological Association. (1988). Trends in ethics cases, common pitfalls, and published resources. *American Psychologist, 43,* 564–572.

Friedman, H. S., & Mitchell, R. R. (2008). Introduction. In H. S. Friedman & R. R. Miitchell (Eds.), *Supervision of sandplay therapy* (pp. 1–9). London: Routledge.

Gutheil, T. G. (1990). Argument for the defendant—expert opinion: Death in hindsight. In R. I. Simon (Ed.), *Review of clinical psychiatry and the law* (pp. 335–339). Washington, D.C.: American Psychiatric Association.

Harrar, W. R., VandeCreek, L., & Knapp, S. (1990). Ethical and legal aspects of clinical supervision. *Professional Psychology: Research and Practice, 21*(1), 37–41.

Helbok, C. M. (2003). The practice of psychology in rural communities: Potential ethical dilemmas. *Ethics & Behaviour, 13*(4), 367–384.

International Society of Sandplay Therapists. (2007). Code of ethics. http://www.isst-society.com/homeng.php?site=ethics. Accessed22 Sept 2014.

Kagle, J. D., & Giebelhausen, P. N. (1994). Dual relationships and professional boundaries. *Social Work, 39,* 213–220.

Lamb, D. H., & Catanzaro, S. J. (1998). Sexual and nonsexual boundary violations involving psychologists, clients, supervisees, and students: Implications for professional practice. *Professional Psychology: Research and Practice, 29*(5), 498–503.

Moleski, S. M., & Kiselica, M. S. (2005). Dual relationships: A continuum ranging from the destructive to the therapeutic. *Journal of Counseling & Development, 83,* 3–11.

Pipes, R. B. (1997). Nonsexual relationships between psychotherapists and their former clients; Obligations of psychologists. *Ethics & Behavior, 7*(1), 27–41.

Pope, K. S. (1989a). Malpractice suits, licensing disciplinary actions, and ethics cases: Frequencies, causes, and costs. *Independent Practitioner, 9*(1), 22–26.

Pope, K. S. (1989b). Therapists who become sexually intimate with a patient: Classification, dynamics, recidivism, and rehabilitation. *Independent Practitioner, 9*(3), 28–34.

Pope, K. S., & Vetter, V. A. (1992). Ethical dilemmas encountered by members of the American Psychological Association. *American Psychologist, 47*(3), 397–411.

Pope, K. S., Tabachnick, B. G., & Keith-Spiegel, P. (1987). Ethics of practice: The beliefs and behaviors of psychologists as therapists. *American Psychologist, 42,* 993–1006.

Psychological Society of Ireland. (2010). Code of professional ethics. Dublin: Psychological Society of Ireland. http://www.psychologicalsociety.ie/find-a-psychologist/PSI%202011-12%20 Code%20of%20Ethics.pdf. Accessed 7 Sept 2014.

Reamer, F. G. (2003). Boundary issues in social work: Managing dual relationships. *Social Work, 48*(1), 121–133.

Recupero, P. R., & Rainey, S. E. (2007). Liability and risk management in outpatient psychotherapy supervision. *Journal of the American Academy of Psychiatry and the Law, 35,* 188–195.

Risley, T. R., & Sheldon-Wildgen, J. (1982). Invited peer review: The AABT experience. *Professional Psychology, 13,* 125–131.

Salisbury, W. A., & Kinnier, R. T. (1996). Posttermination friendship between counselors and clients. *Journal of Counseling & Development, 74,* 495–500.

Sandplay Therapists of America. (2012). Handbook of certified, teaching and practitioner member Requirements and procedures for Sandplay Therapists of America (ISST). http://www.sandplay.org/pdf/STA_Handbook.pdf. Accessed 7 Sept 2014.

Smith, D., & Fitzpatrick, M. (1995). Patient-therapist boundary issues: An integrative review of theory and research. *Professional Psychology: Research and Practice, 26,* 499–506.

Social Workers Registration Board. (2014). Code of conduct guidelines for social workers. Aotearoa, New Zealand: Social Workers Registration Board. http://www.swrb.govt.nz/complaints/code-of-conduct. Accessed 22 Sept 2014.

Stake, J. E., & Oliver, J. (1991). Sexual contact and touching between therapist and client: A survey of psychologists' attitudes and behavior. *Professional Psychology: Research and Practice, 22,* 297–307.

Sterling, D. L. (1992). Practicing rural psychotherapy: Complexity of role and boundary. *Psychotherapy in Private Practice, 10,* 105–127.

Stromberg, C. D., Haggarty, D. J., Leibenluft, R. F., McMillian, M. H., Mishkin, B., Rubin, B. L., & Trilling, H. R. (1988). *The psychologist's legal handbook.* Washington, D.C.: Council for the National Register of Health Service Providers in Psychology.

Sweet, J. J., & Rozensky, R. H. (1991). Professional relations. In M. Hersen, A. E. Kazdin, & A. S. A Bellack (Eds.), *The clinical psychology handbook,* rev ed. (pp. 102–114). New York: Pergamon.

Chapter 7
Special Ethical Considerations in Sandplay Therapy

Sana Loue and Jean Parkinson

7.1 Introduction

This chapter focuses on two issues of critical importance in the context of any therapy, but that assume additional dimensions in the context of sandplay therapy: Client risk associated with a therapeutic modality that has not been specifically identified as an evidence-based practice and termination of services.

It is important to note at the outset that each mental health-related professional organization has its own codes of ethics (Hegeman 2008). In addition to these professional codes of ethics, all mental health professionals are required to adhere to ethics-related provisions of all relevant and applicable statutes, regulations (Hegeman 2008), and court decisions. Professional insurance agreements and employment contacts may delineate yet additional requirements.

7.2 Sandplay Therapy, Risk, and Informed Consent

There has been a significant emphasis in recent years by health insurers, health-care practitioners, and consumers of mental health-care services on the use of evidence-based practice. In psychology, evidence-based practice refers to "the integration of the best available research with clinical expertise in the context of patient characteristics, culture, and preferences" (APA Presidential Task Force on Evidence-Based Practice 2006, p. 273). The definition closely resembles the definition

S. Loue (✉)
Case Western Reserve University, Cleveland, OH, USA
e-mail: sana.loue@case.edu

J. Parkinson
Auckland, New Zealand

© Springer International Publishing Switzerland 2015
S. Loue (ed.), *Ethical Issues in Sandplay Therapy Practice and Research,*
SpringerBriefs in Social Work, DOI 10.1007/978-3-319-14118-3_7

of evidence-based practice in the field of medicine: "the integration of best research evidence with clinical expertise and patient values" (Institute of Medicine 2001, p. 147). The evidence that is used to evaluate any treatment for a specific disorder is assessed with respect to its efficacy—the strength of the causal relationship between the treatment/intervention and the specific disorder—and the clinical utility of the intervention—the generalizability, feasibility, costs, and benefits of the specific intervention (APA Presidential Task Force on Evidence-Based Practice 2006).

The concept of evidence-based practice is related to, but broader than, the concept of empirically supported treatment. First, evidence-based practice considers numerous clinical functions, such as assessment, intervention, and others, whereas the concept of empirically supported treatments focuses on psychological interventions that have been examined in controlled trials and have been found to be effective. Second, evidence-based practice looks at existing research to determine what can be used to obtain the best outcome for a particular patient. That body of research includes the results obtained through clinical observation, qualitative research, systematic case studies, single-case experimental designs, public health research, ethnographic research, process–outcome studies, intervention studies, randomized clinical trials, and meta-analyses. The strength and limitations of the evidence obtained from each of these types of evidence must be evaluated to reach a conclusion. The concept of empirically supported treatments asks whether a specific intervention is effective for a particular diagnosis under specific circumstances (APA Presidential Task Force on Evidence-based practice 2006).

The evaluation of the strength of the available evidence for a particular modality for use with a specific diagnosis may vary across panels of professionals (Feinstein and Horwitz 1997). Additionally, the evaluation of the research evidence does not consider the priorities and preferences of the individual client, the feasibility of using the particular intervention with a specific individual, or the responsiveness of the client to an intervention over time (APA Presidential Task Force on Evidence-Based Practice 2006; Feinstein and Horwitz 1997).

The Australian Psychological Society's (2010) literature review of evidence-based psychological interventions in the treatment of mental disorders provides an example of how evidence-based practice might be used. The literature review evaluates the evidence supporting the use of a particular intervention for specific diagnosis. The source of the research evidence is rated on a scale from I—signifying the systematic review of randomized clinical trials, considered the most reliable form of evidence, to IV—referring to a case series, the least stringent approach to research. The literature review concluded that the use of cognitive behavior therapy for a diagnosis of depression is supported by level I evidence, whereas there is insufficient evidence available to support the use of narrative therapy for dissociative disorders (Australian Psychological Society 2010). This lack of adequate evidence does not mean, however, that narrative therapy is ineffective for the treatment of dissociative disorders; it merely means that this modality has not been subjected to rigorous empirical testing for this disorder. It cannot be assumed to be either effective or ineffective. Clearly, a large "gray zone" exists with respect to some modalities and their use with a variety of disorders (Naylor 1995).

Whether a specific evidence-based practice in psychology is appropriate for a particular client requires the exercise of clinical judgment and an examination of the associated ethical issues. This evaluation may be even more rigorous when contemplating the use of a therapeutic intervention, such as sandplay therapy, for which there is relatively little empirical evidence or only "lower"-level empirical evidence available to establish its efficacy.

As an example, there is relatively strong research evidence supporting the use of cognitive behavior therapy for the treatment of depression in adult clients (Australian Psychological Society 2010). However, this conclusion is based on the "average" client who participated in the studies that constitute the basis of that conclusion (cf. Feinstein and Horwitz 1997). Whether a therapist should use that modality with a particular client is going to depend—or should depend—not only on an understanding of the benefits of that intervention but also the underlying assumptions and the limitations of the intervention; the client's preferences; whether that intervention has been utilized previously with the client and, if it has been, whether its use was successful or counterproductive. The therapist must monitor the client's progress during the course of treatment with this intervention and, if inadequate improvement is forthcoming, adapt the treatment, seek supervision, and/or refer the client for other services or to another therapist. Utilization of cognitive behavior therapy presupposes that a client has the cognitive capacity to engage; is developmentally capable, has a mental health presentation which is accepting of cognitive approaches, and that the therapist is cognizant of new implications in neuropsychotherapy for "talking therapies" (Rossouw 2013).

One difference between a talking therapy such as cognitive behavioral therapy and a largely nonverbal therapy is that sandplay and art psychotherapy are not adjunctive modalities; these are serious psychotherapies which require an understanding of depth psychology. Zoja, a Jungian analyst and sandplay therapist expressed the opinion that

> Sandplay … is the only self-consistent form of therapy: it deals with the pre-verbal and pre-symbolic areas of experience by way of shaping, and manipulation of concrete objects. The hands assume the leading role; the body assumes the leading role. Not narrative, not language. (Zoja 2004, p. 19)

A second difference is referred to in sandplay literature as the triangular constellation/relationship, i.e., the client not only interacts with the therapist but also with the sand. Both participants have the opportunity to be observers of the sand image, which becomes an integral end "product" of the work. It is therefore not only the interactions and process in the sandplay session to which we pay attention but also the image and the client's imaginings and resonance with the entire living and lived experience. As therapists, our "interpretation" is minimal and delayed. This respect for client-led process is a further stumbling block to clinical interpretation leading to scientific proof that sandplay therapy is effective.

The therapist must also consider the ethical questions that may arise in contemplating the use of cognitive behavior therapy for the particular client. The ethical principle of respect for persons requires recognition of the client's autonomy. This is realized through the informed consent process. In turn, the concept of informed consent suggests that the client must be apprised of the risks and benefits of using this therapeutic approach. The ethical principle of beneficence requires that the therapist seeks to minimize potential harm to the client from the use of the intervention, which suggests that a competent, ethical therapist will refrain from utilizing a therapeutic modality that may result in harm to the client. Accordingly, the therapist must be familiar with the existing research pertaining to a specific therapeutic modality and its potential risks and benefits for specific client subgroups, e.g., clients with depression.

For example, the research that has been conducted surrounding the use of cognitive behavior therapy for depression may provide many of the answers needed to address these ethical issues. Extensive research findings derived from studies using a variety of research designs provide much information about the potential benefits, limitations, and risks associated with cognitive behavior therapy as an intervention for depression (Craigie and Nathan 2009; Fava et al. 2004; Kessler et al. 2009; Mohr et al. 2005; Oei and Dingle 2008; Wiles et al. 2008). The competent, ethical therapist would be familiar with this research and would assess the relevance and applicability of the information to the client.

This is not, however, the case with respect to the use of sandplay therapy. Although some research has been conducted to evaluate the reliability, efficacy, effectiveness of sandplay therapy (Fujii 1979; Kamp and Kessler 1970; Miller 1982), the majority of the research that has been conducted consists of case reports that follow clients over time (Enns and Kasaim 2003). As a methodology, case reports are inadequate for the evaluation of efficacy and effectiveness at a group level. In part, this relative scarcity of noncase report research is due to the significant difference that exists between the methodologies and research for verbal therapies compared with sandplay and art therapy, which incorporate nonverbal psychotherapy. Sandplay focuses on the relationship: therapist–client, client–sandplay, and to the process of the client. Weinrib (1983, p. 77) emphasizes the role of self-healing in sandplay: "A basic postulate of sandplay therapy is that deep in the unconscious there is an autonomous tendency, given the proper conditions, for the psyche to heal itself."

The case studies of sandplay have often focused on clients' efforts to resolve specific issues, reduce problematic behaviors, or ameliorate depression (Ammann 1991; Kalff 1980; Weinrib 1983). One of the few studies conducted to utilize existing, validated instruments to assess outcomes over time had significant limitations in the study design including the absence of a control group, a relatively short follow-up period, and uncontrolled confounding (Hong 2011).

The fact that sandplay is not an evidence-based practice does not mean, however, that it is ineffective or should not be used. The research that has been done on sandplay therapy suggests that sandplay can be used successfully with some clients under specific circumstances for particular conditions. This creates greater confidence that sandplay is an effective intervention.

As an example of how this evaluation might be conducted, consider that conventional, anecdotally premised wisdom within the sandplay therapy community holds that sandplay therapy should not be used with clients who have a diagnosis of schizophrenia. One might examine in finer detail what it is about schizophrenia that might contraindicate the use of sandplay. Such features of the illness likely include hallucinations, delusions, dissociation, and disordered thinking. The same features are variously present in other mental illnesses: depression, depression with psychotic features, bipolar disorder, and posttraumatic stress disorder. This evaluation suggests that it is not the diagnosis of schizophrenia per se that mitigates against the use of sandplay therapy, but rather specific features of the illness. A therapist might reasonably conclude that the potential benefits of sandplay therapy would outweigh the potential risks to a client with schizophrenia under specific conditions: (1) the client is not actively psychotic, (2) the client is adhering to an effective medication regimen for schizophrenia, and (3) the therapist can recognize when a client is beginning to dissociate and has the requisite skills and knowledge to intervene effectively.

Several experienced sandplay therapists have commented on the use of sandplay therapy with clients with other, specified disorders. Merlino has utilized sandplay therapy for clients with substance use disorders in order to provide an alternative to verbal therapy. This may have been particularly important for these clients, who were "notoriously recalcitrant to psychotherapy" (Merlino 2004, p. 207). Psychiatrist La Spina has advised against the use of sandplay therapy in the treatment of borderline personality due to the complexity of the problems of transference and countertransference that arise within the patient–therapist dyad of preverbal and nonverbal levels of communication. These issues present themselves as among the most profound, and the therapist must be especially well trained and skilled to be able to address them (La Spina 2004). Bignamini (2004, p. 201) is of the opinion that sandplay offers psychotic patients "the opportunity to access a channel of nonverbal communication where cognitive distortions and emotional conflicts can be expressed symbolically." Sandplay therapy, he believes, provides a "road to the integration of parts which would be otherwise difficult to reach in the course of psychotherapy" (Bignamini 2004, p. 201).

In my clinical experience (Parkinson), clients with low verbal or auditory processing capacity, clients who are affected by trauma and some children and adolescents, particularly those diagnosed with attention deficit disorder, anxiety, selective mutism and Aspergers' characteristics, present challenges for engaging in cognitive therapies. There is a place for visual and experiential modalities to assist as an adjunct in treatment. Frequently, clients in my practice find a relief in the visual, spatial, kinesthetic experience of sand and art materials to express what cannot be put into words. Neuroscience and neuropsychology increasingly support nonverbal therapies for clients with trauma.

Reflective and constructive feedback from other professionals endorsed a presentation made to a child trauma conference (Parkinson and Maoate 2012). Based on parent and child feedback, we summarized their perceived effectiveness of processing trauma and facilitating neural de-escalation through play, sandplay, and art with children living through 2 years of continuous earthquakes and aftershocks in

Christchurch, New Zealand. In a secure, respectful relationship with the children, we reflected and mirrored back emotions expressed through these modalities. The portable therapy room provided a safe, protected space for children to experience de-escalation, and for their parents to learn neurological research-based and developmentally appropriate strategies to facilitate safety and lessen anxiety. Montecchi wrote of sandplay with traumatized clients:

> In handling the sand and positioning the objects within the sandbox to represent a scene, the patients' hands give form to problems that threaten them, to conflicts in their inner world, to real facts that cannot be revealed by words; but they also express the life project concealed within the folds of emotional distress. (Montecchi 2004, p. 130)

In respect of evidence-based structures, to what part of this process would we attribute success? The challenge for us is to find ways to conform to structures demanded by evidence-based practice, while prioritizing the needs of our clients and respecting their individual processes.

In the context of evidence-based practice, Zoja insists that working with the sand is more than a technical operation, which once mastered can be expected to guarantee certain results: "The existential confrontation between the ego and the unconscious is always unpredictable" (Zoja 2004, p. 12). This opinion supports the preponderance of case studies in psychotherapy, particularly art psychotherapy and sandplay therapy. Bosio recognizes it is not always possible to witness a profound renewal of personality: "We have come to feel that the therapeutic efficacy of sandplay can only be seen in the cases where the analyst and patient have recreated a condition of basic trust" (Bosio 2004, p. 177).

It is important to recognize that the evaluation of risk, whether in the example above or in any situation, is essentially an interpretive function that is conducted from a position of privilege. That privilege is not only inherent in the therapist's role as therapist but also may be a function of race, ethnicity, sex, socioeconomic status, and/or educational level. In short, it is a function of the power differential that always exists between therapist and client and that may be enlarged and confounded by personal and contextual factors. An ethical, competent therapist will approach each such evaluation of risk with a sense of humility, understanding that he or she cannot unilaterally make such decisions for the client. Instead, understanding that the therapist's knowledge is imperfect, the therapist and the client can assess the risk and benefit together. This mutual evaluative process may help the therapist to better understand the client and his or her values and may well strengthen the therapeutic alliance.

Clearly, it is critical that a client choosing a therapist or a therapist taking on a new client must consider and evaluate the experience and competence of the therapist, the adequacy and availability of supervisory/consultative support, and the intended modality/modalities. Sandplay practitioners frequently have primary training in other disciplines, such as psychology or social work, and therapeutic modalities, such as Jungian analysis, cognitive behavior therapy, and art therapy. These may underpin but do not replace the need for specific training, personal process, and regular supervision/consultation in sandplay therapy.

7.3 Initiating and Terminating the Therapeutic Relationship

7.3.1 Intermittent or Compressed Therapy

Sandplay therapy is generally considered to be a longer-term, in-depth therapeutic process (Reece 2008). Nevertheless, in practice, there are instances where therapy may be either intermittent or compressed out of necessity. Clients may attend therapy sessions only sporadically due to economic or environmental constraints (Loue 2010) or for an intense, but relatively, short period of time, such as several hours a day for a week, due to illness, employment demands, or family obligations (Parkinson 2013). Research has shown that complications such as financing, employment, and transportation difficulties can result in the unanticipated termination of therapy (Barnett et al. 2000). These circumstances raise several ethical questions in the context of initiating and terminating therapy: (1) what are the risks to the client of utilizing sandplay therapy under such conditions and can such risks be ethically justified and (2) whether and how a client's sense of being abandoned can be avoided, when sandplay is not the longer-term process that one might desire for a client.

> **Case Example 1: Compressed Therapy**
>
> A client living a significant distance from any sandplay practitioners is diagnosed with a serious health condition. Prior knowledge of the effectiveness of sandplay and the perception that verbal therapy will not be effective in her circumstances guides her decision to travel to an experienced trauma practitioner. The client contracts to come daily over a period of 12 days, which is all the time she can afford away from medical assistance. Deep and cathartic work is undertaken during this period. The client hopes to return at a later date to consolidate the therapy but her deteriorating condition precludes travel.

> **Case Example 2: Intermittent Therapy**
>
> A client engaged in a creative field has significant family financial responsibilities. As a result, even at a reduced rate for therapy, he is able to commit financially to only one session every other month. The client is self-motivated to reflect on the work through art and creative journaling in the period between sessions. His capacity for insight and self-reflection is evident in his processing.

Case Example 3: Intermittent Therapy

A child in care can access 6–10 therapy sessions at a time through an agency contract. The child has previously worked with the sandplay therapist and asks to resume therapy. Because of the relationship of trust, the contract is resumed and completed. Subsequently, due to the unstable relationship with her birth family and other traumatizing events during family contact and care, she returns for a block of sessions after each incident. In this case, a new agency contract is required to be negotiated each time. This intermittent therapy arrangement continues for over 5 years.

Despite the conventional wisdom that sandplay therapy must be a long and continuous process, Bradway, one of the foremost early sandplay therapists, observed:

> We are increasingly finding that a sandplay series remains cohesive whether it is brief or scattered over a long period of time. (Bradway and McCoard 1983, p. 173)

7.3.2   Termination and Abandonment

Termination of therapy is

> an intentional process that occurs over time when a client has achieved most of the goals of treatment, and/or when psychotherapy must end for other reasons. (Vasquez et al. 2008, p. 654)

Termination is also warranted when the client is no longer likely to benefit from the therapy or when the client is actually being harmed by the therapy (American Psychological Association 2010; Vasquez et al. 2008). "Other reasons" may include a relocation of the client or the therapist, chronic illness of the therapist or client, or the therapist's retirement (Vasquez et al. 2008). A client may feel abandoned when termination is unexpected or unplanned, or if the client has not had an adequate opportunity to review his or her goals and progress and achieve closure with the therapist.

This same dilemma is illustrated by the following case example:

Case Example 4

A school counselor is providing sandplay therapy to a child in an upper grade in elementary school. Based on teacher reports, the child appears to be benefiting from the therapy. He is instigating fights less frequently, is becoming more communicative verbally, and is focusing more on classroom assignments. State law provides that counseling services cannot be provided to a child through the school counseling service for a period longer than 90 days without receiving written parental consent to continue the services. The therapist has sent several requests for permission to continue these services to the child's one parent, but has not received any response. She has tried phoning the parent, but has not received any return calls. The end of the 90-day period is approaching. She is concerned that the child is not ready to terminate therapy and will feel abandoned if it is terminated at this time. She feels that she cannot ethically terminate services but that she is legally compelled to do so.

One can debate whether the sandplay therapist-school counselor should have begun providing services to the child prior to obtaining written parental consent. Many jurisdictions, however, expressly allow mental health professionals to provide counseling services to minors without parental permission for specific periods of time. For example, the state of Washington permits a child who is 13 years of age or older to consent to receive outpatient or inpatient mental health care without parental consent. Parental/guardian notification is required if the treatment is to be provided on an inpatient basis. (Annotated Revised Code Washington 2012; Columbia Legal Services, Public Health Seattle & King County, & University of Washington Medicine n.d.). In Connecticut, a licensed psychiatrist, psychologist, certified independent social worker, or marital and family counselor is permitted to provide up to six outpatient mental health treatment sessions to a minor child without parental consent or notification if the requirement of notification or consent would cause the child to reject the treatment; treatment is clinically indicated; the failure to provide the treatment would be detrimental to the well-being of the child; the child knowingly and voluntarily sought treatment; and in the opinion of the mental health professional, the child is mature enough to participate in the treatment productively (Connecticut General Statutes 2012). After the sixth session, the provider must notify the child that parent or guardian notification or involvement is required in order to continue with treatment, unless doing so "would be seriously detrimental to the minor's well-being," which must be documented in the record and reviewed and re-documented every sixth session.

Prior to reaching this turning point, the therapist could have utilized various strategies to prepare the child in the event that parental consent could not be obtained, for whatever reason. The therapist might have explained to the child that therapy would only be available for a prescribed period of time (90 days) and that any continuation after that would require additional permission. Reminders of this could be provided periodically to the child. Unfortunately, it is unknown whether such advisories would preclude the establishment of a strong therapeutic alliance; the child might always be waiting for the proverbial shoe to drop and therapy to be terminated.

The therapist might in advance of the end of the 90-day period confer with legal counsel for the school district to review possible courses of action. Could the therapy be interrupted for a brief period of time and then be resumed without contravening state law? Is the parent's de facto refusal to sign the consent form reflective of medical negligence that, if reported to the state, would enable continuing therapy to proceed? Is there any exception under state statute for consent from an alternative person or can the statute be interpreted to require only that "reasonable efforts" be made to obtain written parental consent? Does the fact that the parent is aware of the ongoing therapeutic relationship and has not objected mitigate any potential liability of the school district and the therapist? This situation clearly demonstrates the interwoven challenges as the therapist tries to fulfill her ethical and legal obligations simultaneously.

Assuming that the sandplay therapist cannot continue her work with the child, she can utilize several strategies to ease the termination process. The ending phase of the relationship provides the therapist and the child with an opportunity to discuss together the child's strengths and the positive changes that he has made (Pipes and Davenport 1999).

With adult clients, the termination process can be explored at the initiation of therapy, during the informed-consent process. Circumstances beyond the control of either the therapist or the client may dictate when therapy must end, e.g., restrictions on cost imposed by the terms of the client's health-care insurer or refusal of the health-care insurer to authorize reimbursement for continued services (Vasquez et al. 2008).

References

American Psychological Association. (2010). Ethical principles of psychologists and code of conduct. Washington, DC: American Psychological Association. http://www.apa.org/ethics/code/principles.pdf. Accessed 7 Sept 2014.

Ammann, R. (1991). *Healing and transformation in sandplay*. LaSalle: Open Court.

Annotated Revised Code Washington. § 71.34.530 (2012).

APA Presidential Task Force on Evidence-Based Practice. (2006). Evidence-based practice in psychology. *American Psychologist, 61*(4), 271–285.

Australian Psychological Society. (2010). Evidence-based psychological interventions in the treatment of mental disorders: A literature review (3rd ed.). Melbourne: Australian Psychological Society Limited. https://www.psychology.org.au/Assets/Files/Evidence-Based-Psychological-Interventions.pdf. Accessed 9 Sept 2014.

Barnett, J. E., MacGlashan, S., & Clarke, A. J. (2000). Risk management and ethical issues regarding termination and abandonment. In L. VandeCreek & T. Jackson (Eds.), *Innovations in clinical practice* (pp. 231–246). Sarasota: Professional Resources.

Bignamini, L. (2004). Sacrifice as death and rebirth in adolescent development. In E. P. Zoja (Ed.), *Sandplay therapy: Treatment of psychopathologies* (trans. H. Martin) (pp. 181–202). Switzerland: Daimon Verlag.

Bosio, W. (2004). Image and the analytical relationship in sandplay therapy. In E. P. Zoja (Ed.), *Sandplay therapy: Treatment of psychopathologies* (trans. H. Martin) (pp. 149–179). Switzerland: Daimon Verlag.

Bradway, K., & McCoard, B. (1983). *Sandplay: Silent working of the psyche*. London: Routledge.

Columbia Legal Services, Public Health Seattle & King County, & University of Washington Medicine. (n.d.). Providing health care to minors under Washington law. http://www.washingtonlawhelp.org/   files/C9D2EA3F-0350-D9AF-ACAE-BF37E9BC9FFA/attachments/3924E318-9C45-4556-F8C9-89368DAE1E74/216941minors_health_care_rights.pdf. Accessed 1 Feb 2013.

Connecticut General Statutes. § 19a–14c (2012).

Craigie, M. A., & Nathan, P. (2009). A nonrandomized effectiveness comparison of broad-spectrum group CBT to individual CBT for depressed outpatients in a community mental health setting. *Behavior Therapy, 40*, 302–314.

Enns, C., & Kasaim M. (2003). Hakoniwa: Japanese sandplay therapy. *The Counseling Psychologist, 31*(1), 93–112.

Fava, G. A., Ruini, C., Rafanelli, C., Finos, L., Conti, S., & Grandi, S. (2004). Six-year outcome of cognitive behavior therapy for recurrent depression. *American Journal of Psychiatry, 161,* 1872–1876.

Feinstein, A. R., & Horwitz, R. I. (1997). Problems in the "evidence" of "evidence-based medicine". *American Journal of Medicine, 103,* 529–535.

Fujii, S. (1979). Test-retest reliability of the sand play technique: First report. *British Journal of Projective Psychology and Personality Study, 24,* 21–25.

Hegeman, G. (2008). Ethical dilemmas in sandplay supervision. In H. S. Friedman & R. R. Mitchell (Eds.), *Supervision of sandplay therapy* (pp. 67–72). London: Routledge.

Hong, G. L. (2011). *Sandplay therapy: Research and practice.* New York: Routledge.

Institute of Medicine. (2001). *Crossing the quality chasm: A new health system for the 21st century.* Washington, DC: National Academies.

Kalff, D. M. (1980). *Sandplay: A psychotherapeutic approach to the psyche.* Boston: Sigo.

Kamp, L. N. J., & Kessler, E. S. (1970). The world test: Developmental aspects of a play technique. *Journal of Child Psychology and Psychiatry, 11,* 81–108.

Kessler, D., Lewis, G., Wiles, N., King, M., Welch, S., Sharp, D. J., et al. (2009). Therapist-delivered internet psychotherapy for depression in primary care: A randomised controlled trial. *The Lancet, 374,* 628–634.

La Spina, V. (2004). Sandplay therapy in the treatment of borderline personality disorder. In E.P. Zoja (Ed.). *Sandplay therapy: Treatment of Psychopathologies* (pp. 31–58). Einsiedeln, Switzerland: Daimon Verlag.

Loue, S. (2010). Sandplay with inner city minority adults. Presented at the Annual Meeting of the Sandplay Therapists of America, Boulder, Colorado, June.

Merlino, M. (2004). Images of time: New Departures as a public drug addiction clinic. In E. P. Zoja (Ed.), *Sandplay therapy: Treatment of psychopathologies* (trans. H. Martin) (pp. 203–224). Switzerland: Daimon Verlag.

Miller, R. R. (1982). Investigation of a psychotherapeutic tool for adults: The sand tray. Doctoral dissertation, California School of Professional Psychology, Fresno. Dissertation Abstracts International 43(1–B): 257. University Microfilms Number 82–07557.

Mohr, D. C., Hart, S. L., Julian, L., Homos-Webb, L., Vella, L., et al. (2005). Telephone-administered psychotherapy for depression. *Archives of General Psychiatry, 62,* 1007–1014.

Montecchi, F. (2004). The self and family archetypes in children. In E. P. Zoja (Ed.), *Sandplay therapy: Treatment of psychopathologies* (trans. H. Martin) (pp. 107–148). Switzerland: Daimon Verlag.

Naylor, C. D. (1995). Grey zones of clinical practice: Some limits to evidence-based medicine. *The Lancet, 345,* 840–842.

Oei, T. P. S., & Dingle, G. (2008). The effectiveness of group cognitive behaviour therapy for unipolar depressive disorders. *Journal of Affective Disorders, 107,* 5–21.

Parkinson, J. (2013). Case presentation. Presented at Caring for the Soul, Einsiedeln, Switzerland, July.

Parkinson, J. & Maoate, A. (2012). After the earthquakes it was muted. Presentation to the First Australasian Child Trauma Conference, Gold Coast, Australia.

Pipes, R.B., & Davenport, R.S. (1999). *Introduction to psychotherapy: Common clinical wisdom.* (2nd ed.). Boston, MA: Allyn & Bacon.

Reece, S. T. (2008). Sandplay supervision in a community mental health center. In H. S. Friedman & R. R. Mitchell (Eds.), *Supervision of sandplay therapy* (pp. 113–135). New York; Routledge.

Rossouw, P. (2013). The neuroscience of talking therapies. Neuropsychotherapy in Australia, 24, Nov–Dec.

Vasquez, M. J. T., Bingham, R. P., & Barnett, J. E. (2008). Psychotherapy termination: Clinical and ethical responsibilities. *Journal of Clinical Psychology: In Session, 64*(5), 653–665.

Weinrib, E. L. (1983). *Images of the self: The sandplay therapy process*. Cloverdale: Temenos.

Wiles, N. J., Hollinghurst, S., Mason, V., & Musa, M. (2008). A randomized controlled trial of cognitive behavioural therapy as an adjunct to pharmacotherapy in primary care based patients with treatment resistant depression: A pilot study. *Behavioural and Cognitive Psychotherapy, 26,* 21–33.

Zoja, E. P. (2004). Understanding with the hands. In E. P. Zoja (Ed.), *Sandplay therapy: Treatment of psychopathologies* (trans. H. Martin) (pp. 13–29). Switzerland: Daimon Verlag.

Index

© Springer International Publishing Switzerland 2015   99
S. Loue (ed.), *Ethical Issues in Sandplay Therapy Practice and Research,*
SpringerBriefs in Social Work, DOI 10.1007/978-3-319-14118-3